Dynamically Coupled Rigid Body-Fluid Flow Systems

Banavara N. Shashikanth

Dynamically Coupled Rigid Body-Fluid Flow Systems

Springer

Banavara N. Shashikanth
Mechanical and Aerospace Engineering
New Mexico State University
Las Cruces
NM, USA

ISBN 978-3-030-82645-1 ISBN 978-3-030-82646-8 (eBook)
https://doi.org/10.1007/978-3-030-82646-8

Mathematics Subject Classification: 76B47, 76M60, 37K05, 37J15, 53D20, 37N05, 70H03, 70H05, 76B07, 97M50

This Springer imprint is published by the registered company Springer Nature Switzerland AG
The registered company address is: Gewerbestrasse 11, 6330 Cham, Switzerland

Preface

The time-honored method of studying the dynamics of fluid-solid interaction problems has been to either fix the solid or to prescribe its motion, for example, a steady translation or a regular oscillation. Investigations then focus primarily on the dynamics of the fluid flow and the study of the various features that develop, such as, for example, boundary layers, vortices, shock waves, and turbulence. An equally important focus, especially in engineering applications, is on computing the fluid dynamic loads exerted on the solid with a view to designing it to withstand these loads as it executes the prescribed motions—some familiar engineering examples being the aerodynamic loads on an aircraft and the hydrodynamic loads on a submarine. The dynamics of solids—rigid and deformable, in the absence of any ambient fluid—is also a well-researched topic. In each of these two principle branches of mechanics, many fundamental results have been obtained and many fundamental principles elucidated, starting from the very birth of the subject of theoretical mechanics.

Relatively far less attention has been paid to the dynamics of fluid-solid interaction problems where the solid is not constrained to stay put or move in a prescribed manner, and is free to move in response to the instantaneous fluid stress field on its surface. The focus in such a setting is as much on the dynamics of the solid as on the dynamics of the fluid. Moreover, the motion of the solid in turn influences the flow of the fluid around it, thereby intricately coupling the fluid and the solid dynamics. This is the definition of the dynamically coupled setting, and the topic, in the author's opinion, merits consideration as an independent branch of theoretical mechanics.

Despite the fact that Kirchhoff investigated perhaps the first problem in this setting as far back as the nineteenth century, and despite the fact that there are examples of such interactions all around us in nature—objects lifted up by a strong wind, objects tossed in or floating on water—the topic did not attract much research interest for a long time. The reasons are not difficult to understand. From a mathematical point of view, coupled dynamics is invariably nonlinear. In addition to the complex, inherent nonlinearities of a fluid flow, one must now also deal with the nonlinear response of the body, and the dynamic coupling. The topic therefore remained more or less dormant till the birth and development of the subjects of dynamical systems and nonlinear dynamics. Theoretical investigations in the meanwhile were mainly restricted to cases where linearization techniques could be used, namely in the study of a solid dynamically coupled with water waves ("floating objects") whose equations are amenable to linearization. The case of a solid dynamically coupled with vortices, whose equations are not linearizable, had to wait for almost a 120 years after Kirchhoff's work before it was investigated. And from the point of view of applications, traditional engineering vehicles designed for locomotion in a fluid environment (aircraft, ships, submarines, etc.) rarely required knowledge of the dynamics of an uncontrolled solid object.

A third reason may be postulated as to why the subject did not garner much attention. The fact is dynamically coupled systems with *buoyant* solid objects fully immersed in a fluid environment have been studied, though again not so much the nonlinearities. This is essentially the subject of bodies falling or rising freely in a fluid environment. The notion of a body free to respond to fluid stresses was perhaps more often associated with such problems. Here gravity typically is the dominant feature, and even when interesting dynamical features are seen, such as, for example, the deformations of rising bubbles or the oscillations of a falling sheet of paper, gravity is an essential ingredient of such dynamics. For a long time, there was probably not enough motivation, theoretically or in the real world, to consider the case of neutrally buoyant bodies fully immersed in a fluid environment.

The subject of dynamical systems/nonlinear systems has grown enormously in the past four decades or so, while at the same time there has been development of new "green" paradigms for the design of engineering devices dynamically interacting with fluids. For example, underwater biomimetic vehicles that typically have limited propulsive power, unlike traditional vehicles, and need to maneuver optimally in the fluid environment. Understanding of the coupled dynamics of fluid and vehicle therefore becomes paramount. There are also other engineering examples, where the dynamic coupling is partial, for example, floating offshore wind turbines, and it is important when dealing with such problems to have a knowledge of the fully dynamically coupled problem.

Theoretical investigations on the topic of dynamically coupled systems therefore have been growing at a slow but steady rate. It seems timely and appropriate to present a compilation of important work done so far in the area, and to present this in a framework whose foundations are Kirchhoff's equations. This also allows bringing together under a common banner topics which have traditionally been studied in separate research groups, such as, for example, the system of a rigid body and waves, and the system of a rigid body and vortices. The focus is on Lagrangian and Hamiltonian methods. Functional analytic approaches that investigate existence and smoothness of solutions in Eulerian and Navier-Stokes settings are not covered.[1] Following an introduction to Kirchhoff's equations of motion, the book discusses several extensions of Kirchhoff's work, with the focus on the extensions involving

[1] The interested reader may look up journals such as, for example, *Journal of Mathematical Fluid Mechanics* and *Archive for Rational Mechanics and Analysis* for papers that use such approaches, and references therein.

vortices. The equations of motions of these systems and their Lagrangian and Hamiltonian formulations are presented and discussed. In some cases, detailed derivations are included, especially on the Euler-Lagrange equations for some of the models which, to the best of my knowledge, have not been published before. In addition, wherever appropriate, I have also discussed applications.

In writing this book, I have tried to strike a balance between the modern theories of reduction of Lagrangian and Hamiltonian systems with symmetry—or geometric mechanics, as the subject is more commonly known—and classical fluid mechanics. I have not written this as a purely geometric mechanics book and the reasons are twofold: firstly, there remains much to be done in applying these very technical theories to what is essentially a difficult problem, especially the extensions discussed in the later chapters, and these are tasks not easily done single handedly. Secondly, I want to emphasize that this topic is as much about applications as about theory, and it is for this reason that one chapter of this book is devoted to experimental and numerical work done by my collaborators.

My intention is to make the book understandable to anyone with basic knowledge of Lagrangian and Hamiltonian mechanics, and of theoretical fluid mechanics—regardless of whether they are mathematicians, physicists, or engineers. To aid the reader unfamiliar with the subject, an appendix which briefly introduces basic concepts and terminology in geometric mechanics has been included. I believe the book is written at a level which would make it comprehensible to an advanced doctoral candidate with this background knowledge.

Finally, it goes without saying that this is an ongoing project. From the point of view of applications such as, for example, biomimetic locomotion, there are at least two more sequels that one could think of: one on the topic of dynamically coupled elastic bodies and fluids, and the second on the topic of adding control actuation to these models and developing control theoretic models.

Las Cruces, NM, USA
May 2021

Banavara Shashikanth

Contents

Chapter 1
Kirchhoff's Insufficiently-Celebrated Equations of Motion

1.1 Introduction

Gustav Robert Kirchhoff (1824–1887) made pioneering contributions to physics and engineering. His most famous contributions were in electrical circuits, black body radiation, elasticity, and spectroscopy. In his biographical sketch on Kirchhoff, P. G. Tait writes [161]:

> While there are nowadays hundreds of men thoroughly qualified to work out, to its details, a problem already couched in symbols, there are but few who have the gift of putting an entirely new physical question into such a form. The names of Stokes, Thomson, and Clerk-Maxwell will at once occur to British readers as instances of men possessing such power in a marked degree. Kirchhoff had in this respect no superior in Germany, except his lifelong friend and colleague v. Helmholtz.

His contributions to fluid mechanics are perhaps not as well-known. Indeed most graduate level fluid mechanics textbooks, with the obvious exceptions of Lamb [92] and Milne-Thomson [117], contain no description of the equations of motion that he derived in his paper of 1869 [81]. It is therefore appropriate that the first chapter of this book, in which these equations form a central theme, be devoted to them.

His equations are first described in their classical form and then in the framework of modern Lagrangian and Hamiltonian mechanics. The reader is referred to the excellent text by Marsden and Ratiu [109] for many of the geometric mechanics and variational tools used in this chapter and the remainder of the book. A brief introduction to these notions is also presented in Appendix A.

B. N. Shashikanth, *Dynamically Coupled Rigid Body-Fluid Flow Systems*,
https://doi.org/10.1007/978-3-030-82646-8_1

1.2 Kirchhoff's Equations

Kirchhoff considered the case of a neutrally buoyant rigid body B, with boundary ∂B, immersed in an incompressible fluid[1] occupying a domain $D \subset \mathbb{R}^3$ that extends to infinity in all directions away from the body, as shown schematically in Fig. 1.1. If the body, or the fluid, is set into motion initially, the motion of the body sets

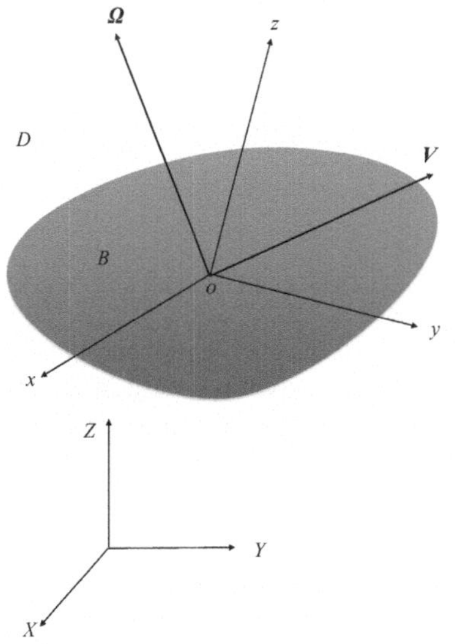

Fig. 1.1 A neutrally buoyant rigid body B dynamically interacting with an incompressible, inviscid, and irrotational fluid occupying the domain $D \subset \mathbb{R}^3$. The body-fixed frame xyz is centered at o, the body's center of mass, with axes parallel to the principal axes of the body. XYZ is the stationary frame

up a flow of the fluid since the fluid will rush in to fill the gap created by the body ("nature abhors a vacuum"); in turn, the hydrodynamic stresses on the body keep the body in motion. In the absence of any external constraining or controlling force, the body and the fluid will thus dynamically interact. This is the *dynamically coupled problem* considered by Kirchhoff.

In the absence of viscosity and vorticity, he showed that the evolution of the coupled body+fluid system is completely described by the equations

$$\begin{aligned} \frac{d\mathbf{L}}{dt} + \Omega \times \mathbf{L} &= 0 \\ \frac{d\mathbf{A}}{dt} + \Omega \times \mathbf{A} + \mathbf{V} \times \mathbf{L} &= 0, \end{aligned} \tag{1.1}$$

[1] Throughout this book, unless otherwise specified, the value of the uniform density of the fluid will be set equal to unity.

where $\mathbf{V}, \Omega, \mathbf{L}$, and $\mathbf{A}$ are all vectors referred to a Cartesian frame $\{\mathbf{e}_i, \mathbf{e}_2, \mathbf{e}_3\}$, attached to the moving body, whose axes coincide with the principal axes of the body and whose origin is at the body's center of mass.[2] The vectors $\mathbf{V}$ and Ω denote the linear velocity of the body's center of mass and the angular velocity of the body, respectively, and the vectors $\mathbf{L}$ and $\mathbf{A}$ denote the linear and angular momenta,[3] respectively, of the combined body+fluid system. These are given by

$$\mathbf{L} := m\mathbf{V} - \int_{\partial B} \Phi_B \mathbf{n}\, v, \quad \mathbf{A} := \mathbf{I} \cdot \Omega - \int_{\partial B} \mathbf{l} \times \Phi_B \mathbf{n}\, v, \tag{1.2}$$

where m is the mass of the body, $\mathbf{I}$ is its moment of inertia tensor (diagonalized by the choice of the principal-axes frame), $\mathbf{l}$ is the position vector in the body-fixed frame, $\mathbf{n}$ is the unit normal field, and Φ_B is the velocity potential of the fluid flow satisfying the following Neumann problem:

$$\nabla^2 \Phi_B = 0 \text{ in } D, \quad \nabla \Phi_B \cdot \mathbf{n} = \mathbf{U} \cdot \mathbf{n} \text{ on } \partial B, \quad \Phi_B \to 0 \text{ as } \|\, \mathbf{l}\, \| \to \infty, \tag{1.3}$$

where $\mathbf{U} = \mathbf{V} + \Omega \times \mathbf{l}$.

Convention for Unit Normal Fields

Throughout this book, unless otherwise specified, the unit normal vector field on the boundary of a fluid domain will be assumed to be inward pointing. In particular, this means that on a body surface the unit normal field points away from the body and into the fluid. Application of the standard integral theorems of vector calculus, where the unit normals are outward pointing, may therefore entail changes in signs of the body-boundary integrals, as in (1.2).

The brilliant insight obtained by Kirchhoff was that Φ_B can be written as

$$\Phi_B(\mathbf{l}, t) = \sum_{i=1}^{3} \left(V_i(t)\psi_i(\mathbf{l}) + \Omega_i(t)\zeta_i(\mathbf{l})\right), \quad (1,2,3) \equiv (x,y,z), \tag{1.4}$$

where each ψ_i satisfies the Neumann problem

$$\nabla^2 \psi_i = 0 \text{ in } D, \quad \nabla \psi_i \cdot \mathbf{n} = \mathbf{e}_i \cdot \mathbf{n} \text{ on } \partial B, \quad \psi_i \to 0 \text{ as } \|\, \mathbf{l}\, \| \to \infty, \tag{1.5}$$

and each ζ_i satisfies the Neumann problem

$$\nabla^2 \zeta_i = 0 \text{ in } D, \quad \nabla \zeta_i \cdot \mathbf{n} = (\mathbf{e}_i \times \mathbf{l}) \cdot \mathbf{n} \text{ on } \partial B, \quad \zeta_i \to 0 \text{ as } \|\, \mathbf{l}\, \| \to \infty. \tag{1.6}$$

It is easy to see that the Neumann problems solved by these component potential functions correspond to the body translating or rotating with unit linear or angular velocity along each of the frame directions. These time-invariant unit potential

---

[2] Unless otherwise stated, all body-fixed frames in this book will be assumed to satisfy these conditions.

[3] The use of the terminology "momenta" is discussed later.

functions, for a given body-fixed frame, are therefore dependent only on the shape of the body and are determined completely by it.

Moreover, substituting (1.4) into the fluid momenta integrals and using the boundary conditions in (1.5) and (1.6),

$$\begin{pmatrix} -\int_{\partial B} \Phi_B \mathbf{n}\, \nu \\ -\int_{\partial B} \mathbf{l} \times \Phi_B \mathbf{n}\, \nu \end{pmatrix} = \mathbf{M_a} \cdot \begin{pmatrix} \mathbf{V} \\ \Omega \end{pmatrix}, \tag{1.7}$$

where $\mathbf{M_a}$ is a symmetric, positive-definite 6×6 matrix called the *added mass tensor*[4] given by

$$\mathbf{M_a} := - \begin{pmatrix} \int_{\partial B} \psi_1 n_1\, \nu & \int_{\partial B} \psi_2 n_1\, \nu & \int_{\partial B} \psi_3 n_1\, \nu & \cdots\cdots\cdots \\ \int_{\partial B} \psi_1 n_2\, \nu & \int_{\partial B} \psi_2 n_2\, \nu & \int_{\partial B} \psi_3 n_2\, \nu & \cdots\cdots\cdots \\ \int_{\partial B} \psi_1 n_3\, \nu & \int_{\partial B} \psi_2 n_3\, \nu & \int_{\partial B} \psi_3 n_3\, \nu & \cdots\cdots\cdots \\ \int_{\partial B} \psi_1 (\mathbf{l} \times \mathbf{n})_1\, \nu & \int_{\partial B} \psi_2 (\mathbf{l} \times \mathbf{n})_1\, \nu & \int_{\partial B} \psi_3 (\mathbf{l} \times \mathbf{n})_1\, \nu & \cdots\cdots\cdots \\ \int_{\partial B} \psi_1 (\mathbf{l} \times \mathbf{n})_2\, \nu & \int_{\partial B} \psi_2 (\mathbf{l} \times \mathbf{n})_2\, \nu & \int_{\partial B} \psi_3 (\mathbf{l} \times \mathbf{n})_2\, \nu & \cdots\cdots\cdots \\ \int_{\partial B} \psi_1 (\mathbf{l} \times \mathbf{n})_3\, \nu & \int_{\partial B} \psi_2 (\mathbf{l} \times \mathbf{n})_3\, \nu & \int_{\partial B} \psi_3 (\mathbf{l} \times \mathbf{n})_3\, \nu & \cdots\cdots\cdots \end{pmatrix}. \tag{1.8}$$

The entries in the fourth to sixth columns of $\mathbf{M_a}$ are the entries of the first to third columns, respectively, but with ψ_1 replaced with ζ_1, ψ_2 replaced with ζ_2, and ψ_3 replaced with ζ_3. The symmetry is shown using the boundary conditions in (1.5) and (1.6) and invoking the following well-known result, applied to two harmonic functions f and g in $\mathbb{R}^3 \backslash B$:

$$\int_{\partial B} (f \nabla g \cdot \mathbf{n} - g \nabla f \cdot \mathbf{n})\ \nu = \int_{\mathbb{R}^3 \backslash B} (f \nabla^2 g - g \nabla^2 f)\ \mu, \tag{1.9}$$

$$\Rightarrow \int_{\partial B} (f \nabla g \cdot \mathbf{n} - g \nabla f \cdot \mathbf{n})\ \nu = 0.$$

The substitution (1.4) also allows the fluid kinetic energy to be written as a quadratic form in the body velocities:

$$(K.E.)_{\text{fluid}} = \frac{1}{2} \int_D \nabla \Phi_B \cdot \nabla \Phi_B\, \mu = -\frac{1}{2} \int_{\partial B} \Phi_B \nabla \Phi_B \cdot \mathbf{n}\, \mu, \tag{1.10}$$

$$= \frac{1}{2} (\mathbf{V}, \Omega) \cdot \mathbf{M_a} \cdot (\mathbf{V}, \Omega)^T, \tag{1.11}$$

using (1.7) and the boundary condition in (1.3). The decomposition (1.4) therefore allows the dynamical effects of the fluid to be completely expressed via the added mass tensor.

[4] The notion of added mass by itself, however, predates Kirchhoff's work and is independent of the dynamically coupled setting; see later.

Introducing the body mass tensor $\mathbf{M_b}$, the body+fluid momenta and kinetic energy become, respectively,

$$\begin{pmatrix} \mathbf{L} \\ \mathbf{A} \end{pmatrix} = (\mathbf{M_b} + \mathbf{M_a}) \cdot \begin{pmatrix} \mathbf{V} \\ \Omega \end{pmatrix}, \tag{1.12}$$

$$(K.E.)_{\text{solid}} + (K.E.)_{\text{fluid}} = \frac{1}{2} (\mathbf{V}, \Omega) \cdot (\mathbf{M_b} + \mathbf{M_a}) \cdot (\mathbf{V}, \Omega)^T. \tag{1.13}$$

In the last equation, since the left hand side is always a positive function, $\mathbf{M_b} + \mathbf{M_a}$ is positive-definite and therefore invertible. The same comment applies to $\mathbf{M_a}$ by itself, since it is the matrix of the positive function $(K.E.)_{\text{fluid}}$. Equations (1.12) make system (1.1) a closed system in the variables $(\mathbf{L}, \mathbf{A})$, for a given body shape. Inverting the former equations then allows one in principle to obtain $\Phi(\mathbf{l}, t)$, using (1.4), from the integrated trajectories $(\mathbf{L}(t), \mathbf{A}(t))$.

Writing the symmetric mass matrix in block form

$$\mathbf{M_b} + \mathbf{M_a} = \begin{pmatrix} \mathbf{M}_{11} & \mathbf{M}_{12} \\ \mathbf{M}_{12}^T & \mathbf{M}_{22} \end{pmatrix}, \tag{1.14}$$

where $\mathbf{M}_{11}$ and $\mathbf{M}_{22}$ are also symmetric (and positive-definite[5]) matrices, note that

$$\mathbf{L} = \begin{pmatrix} \mathbf{M}_{11} & \mathbf{M}_{12} \end{pmatrix} \cdot \begin{pmatrix} \mathbf{V} \\ \Omega \end{pmatrix} \tag{1.15}$$

and

$$\mathbf{A} = \begin{pmatrix} \mathbf{M}_{12}^T & \mathbf{M}_{22} \end{pmatrix} \cdot \begin{pmatrix} \mathbf{V} \\ \Omega \end{pmatrix}. \tag{1.16}$$

Note that the body's angular velocity can contribute to $\mathbf{L}$ and the body's linear velocity can contribute to $\mathbf{A}$.

Remarks

1. Kirchhoff's equations of course bear a resemblance to Euler's equations for a free rigid body in body-fixed frame [51, 109], the difference being the $\mathbf{V} \times \mathbf{L}$ term. Due to the presence of the added mass tensor this term is not generally equal to zero. As discussed later, the two systems are both governed by the same type of Poisson bracket.
2. The identification of the boundary integrals in (1.2) with the fluid momenta is strictly not true. This is due to the well-known fact in fluid dynamics [15, 147] that the fluid momenta integrals—by definition $\int_D \nabla \Phi_B \, \mu$ and $\int_D \mathbf{l} \times \nabla \Phi_B \, \mu$—may not exactly converge to these values in an externally unbounded domain. In other words, in transforming these domain integrals into boundary integrals,

---

[5] Using the same argument of the positivity of the kinetic energy, restricted to purely translation and purely rotational motions.

in addition to the boundary integrals in (1.2), there are corresponding boundary integrals on a large geometric boundary enclosing the body. As this boundary goes to infinity, these terms do not always go to zero but might go to infinity or, in some cases, even go to a non-zero finite value. However, one of the important messages that come across from Kirchhoff's work is that equations (1.1) can be derived purely from energy principles, the energy in this case being the sum of the body plus fluid kinetic energies. The fluid kinetic energy, given by (1.10), unlike the momenta integrals, *is* a convergent integral. Moreover, even though the fluid momenta at any time instant t may be ill-defined, the time rate of change of the momenta is well-behaved and the far-field terms in Newton's Second Law all go to zero. This is demonstrated, for example, in the model extensions with vortices [156, 157]. For the remainder of the book, the momentum/momenta terminology will continue to be used to refer to quantities like **L** and **A**, which may be part of but not completely equal to the fluid domain integrals.

Decompositions analogous to (1.4) exist for the vector potential field in $\mathbb{R}^3$ and for the streamfunction in $\mathbb{R}^2$. The former will be considered later; in the latter case the streamfunction ψ_B can be written as

$$\psi_B(\mathbf{l}, \mathbf{V}(t), \Omega(t)) = \sum_{i=1}^{2} V_i(t)\eta_i(\mathbf{l}) + \Omega(t)\kappa(\mathbf{l}), \tag{1.17}$$

where the unit streamfunctions (η_1, η_2, κ) are the harmonic conjugates of the unit potentials that appear in (1.4) and satisfy the following boundary value problems:

$$\begin{aligned}
\nabla^2\eta_1 &= 0 \text{ in } D, \quad \nabla\eta_1 \cdot \mathbf{t} = -\mathbf{e}_2 \cdot \mathbf{t} \text{ on } \partial B, \quad \eta_1 \to 0 \text{ as } \|\mathbf{l}\| \to \infty, \\
\nabla^2\eta_2 &= 0 \text{ in } D, \quad \nabla\eta_2 \cdot \mathbf{t} = \mathbf{e}_1 \cdot \mathbf{t} \text{ on } \partial B, \quad \eta_2 \to 0 \text{ as } \|\mathbf{l}\| \to \infty, \\
\nabla^2\kappa &= 0 \text{ in } D, \quad \nabla\kappa \cdot \mathbf{t} = \mathbf{l} \cdot \mathbf{t} \text{ on } \partial B, \quad \kappa \to 0 \text{ as } \|\mathbf{l}\| \to \infty.
\end{aligned}$$

An advantage of using the streamfunction instead of the velocity potential function is that the former has a particularly simple representation on ∂B, regardless of the (smooth) shape of the body [117, 150], viz.

$$\begin{aligned}
\eta_1(\mathbf{l})_{|\partial B} &= -y_b, \\
\eta_2(\mathbf{l})_{|\partial B} &= x_b \\
\kappa(\mathbf{l})_{|\partial B} &= \frac{x_b^2 + y_b^2}{2},
\end{aligned}$$

where $\mathbf{l}_{|\partial B} \equiv (x_b, y_b)$. Other interesting relations can be derived from these and are described in a later chapter.

Translating or/and Rotating Equilibrium Solutions

Write (1.1) in the form

$$(\mathbf{M_b}+\mathbf{M_a})\cdot\left(\frac{d\mathbf{V}}{dt},\frac{d\Omega}{dt}\right)^T = -(\Omega\times\mathbf{L},\Omega\times\mathbf{A}+\mathbf{V}\times\mathbf{L})^T.$$

Since $\mathbf{M_b}+\mathbf{M_a}$ is invertible, it is clear that to obtain solutions corresponding to a steady translation and rotation of the rigid body, given by $\mathbf{V}(t)=\mathbf{V}(0)$ and $\Omega(t)=\Omega(0)$, the following conditions are necessary and sufficient:

$$\Omega(0)\times\mathbf{L}(0)=0,\quad \Omega(0)\times\mathbf{A}(0)+\mathbf{V}(0)\times\mathbf{L}(0)=0.$$

Pure translating equilibria are obtained by requiring, in addition, $\Omega(0)=0$. The above conditions then reduce to

$$\mathbf{V}(0)\times\mathbf{L}(0)=0\Rightarrow\mathbf{V}(0)\times\mathbf{M}_{11}\cdot\mathbf{V}(0)=0.$$

Kirchhoff interpreted the last condition as the condition for $\mathbf{V}(0)$ to be orthogonal with the following ellipsoid defined in $\mathbf{V}(0)$-space:

$$\mathbf{V}(0)\cdot\mathbf{M}_{11}\cdot\mathbf{V}(0)=\text{constant},$$

which holds when $\mathbf{V}(0)$ points along any of the three principal axes of the ellipsoid thereby providing three values of $\mathbf{V}(0)$ corresponding to pure translation.

Some more special cases and their dynamical features were considered by Kirchhoff. It should be noted that Thomson and Tait in their seminal treatise on theoretical mechanics (or "Natural Philosophy") [163] had earlier investigated special cases involving bodies of revolution; indeed their work seems to have been the inspiration behind Kirchhoff's own investigations. In addition, the texts [16, 92, 117] contain a good discussion of these cases. In Birkhoff's text [16] there is whole chapter devoted to the added mass concept and its history. According to him,

> the phenomenon of added mass was first discovered experimentally

by DuBuat in 1776.

1.3 The Legacy of Kirchhoff's Equations

Though the notion of added mass became standard in marine engineering—see, for example, [88][6]—the dynamically coupled method itself, unfortunately, did not become popular in engineering fields. Especially in the fields of marine, aerospace, and

[6] This useful reference discusses analytic tools for computing added mass coefficients for a variety of different geometries and also provides a long bibliography of Russian works on the topic.

mechanical engineering where, traditionally, there have existed numerous applications of bodies moving in fluid environments. The reasons however are not difficult to understand. Typically in all such applications the body is designed to navigate along prescribed trajectories propelled by engines powerful enough to *counter* the fluid stresses, rather than be driven passively by them as in the dynamically coupled setting. For example, in aerospace engineering heavier-than-air aircraft have to generate sufficient forward velocity to generate the necessary lift force while at the same time overcoming the drag force. The design of ships must be able to overcome the destabilizing effect of surface waves.

In addition, there are complicating flow factors, not taken into consideration in Kirchhoff's model. In general, over the entire flow Reynolds number range of applications, there are different kinds of complicating flow features. At low Reynolds numbers, such as in relatively recent micro-fluid and nano-fluid applications, viscosity dominates in the flow domain and Stokes-flow-like models are more appropriate. At intermediate Reynolds numbers, viscosity is confined to thin boundary layers but these can decisively influence the dynamics through phenomenon like vortex shedding. And at high Reynolds number, the phenomenon of turbulence kicks in. At high speeds the compressibility of the fluid introduces a whole additional set of flow features, such as shock waves. Theoretical, experimental, and computational work in these fields have naturally focused on the body being fixed or executing simple prescribed motions with more attention paid to these complex flow features. The scientific literature using such approaches is large.

On the other hand, there are some obvious advantages of the dynamically coupled approach. Computing fluid stresses on the surface of the body and the resultant fluid dynamic forces becomes less important; instead one tracks the motion of the rigid body. Dynamics is replaced by kinematics. Secondly, if there are no fixed (external) boundaries or if these are sufficiently far away, then one has conserved quantities or approximately conserved quantities related to the Euclidean group symmetries of the system. These quantities, as in Kirchhoff's equations, are the system momenta in a spatially-fixed frame. Since these can be changed only by forces external to the entire system, these are conserved even for viscous flows.

Mathematically too, Kirchhoff's equations were not appreciated for a long time. The seeds of the subjects of dynamical systems and the qualitative theory of differential equations were barely being laid around the time his equations were written. A similar statement applies to Lagrangian and Hamiltonian mechanics, which at that time was understood only in a classical sense. Indeed Kirchhoff derived his equations from the classical action principle of Hamilton.[7] The modern differential geometric outlook involving Poisson brackets, symplectic forms, Lie groups, symmetries, reduction, and momentum maps was still far from development and acceptance.

---

[7] The seminal papers and books of Henri Poincaré and Sophus Lie, which laid the foundations of the modern developments of these fields, respectively, were written within two decades of Kirchhoff's paper.

But in the last three decades or so, advances in both applications and in the above mathematical fields have cast Kirchhoff's equations in a new light. Engineers have sought to build small unmanned underwater vehicles, either remotely piloted or functioning as robots, for scientific, environmental, recreational, and military applications. These autonomous underwater vehicles (AUVs) have simple axisymmetric shapes, and others draw inspiration from nature's aquatic swimmers, such as many different kinds of fish. A common feature of these vehicles is the limited amount of propulsive power available. Energy-efficient locomotion and seeking ways of harnessing the fluid's energy and momentum become important, especially for biomimetic and biomechanical machines. The dynamically coupled setting thus becomes very relevant for such problems. Interest has also spiked worldwide in building "green energy" devices like, for example, offshore wind turbines. In particular, floating offshore wind turbines are a relatively recent development in which there is partial dynamical coupling between the solid and the fluid. In these turbines there is a complex interaction of the solid structure, water waves, and both air and wind vortices, and the extensions of Kirchhoff's equations developed in later chapters become relevant. The transport of environmentally degrading passive objects by rivers and seas is another potential area of applications.

On the mathematical side, from the modern dynamical systems perspective, several important features of Kirchhoff's equations are now recognized. Firstly, it is a system of nonlinear ODEs, and though the nonlinearities in the equations are algebraic, every student of dynamical systems knows not to underestimate the complexity of behavior of a system with algebraic nonlinearities [102]. Kirchhoff's equations are also a rare example of an *exact finite-dimensional* representation of a fluid flow.[8] The phase space of a fluid flow—even under the incompressible, inviscid, and irrotational assumptions—is, generically, infinite-dimensional. More importantly, from the perspective of modern Hamiltonian systems, Kirchhoff's equations are a *Lie–Poisson* system, i.e. a Hamiltonian system governed by the Lie–Poisson bracket, a non-canonical Poisson bracket. These brackets are obtained by a process of symmetry reduction and Kirchhoff's equations are indeed an example of a symmetry reduced system, as discussed later in the chapter. The differential geometric tools of modern Hamiltonian mechanics, introduced in Appendix A, and other geometric and topological methods in dynamical systems can be applied to these equations. Moreover, with relevance to the applications mentioned, control theoretic ideas, especially geometric control and optimal control ideas, can also be developed. As a simple example, an external control force through the body's center of mass or an external control torque results in a control model in which these terms appear affinely in the equations [95].

[8] With the caveat that for bodies of non-standard shapes, obtaining the component potential functions (1.5) and (1.6) may require solving the corresponding infinite-dimensional Laplace equation.

1.4 The Geometric Mechanics of Kirchhoff's Equations

Arnold[9] in his pioneering papers [8, 10] showed that the following two systems (considered separately): (a) the flow of an incompressible, inviscid fluid (in an unbounded domain or in a domain with stationary boundaries) in an Eulerian or spatial representation and (b) the motion of a free rigid body represented in a body-fixed frame, are both examples of symmetry reduced Hamiltonian systems. In (a), the symmetry group $G = \mathrm{Diff}_{\mathrm{vol}}(D)$, the infinite-dimensional Lie group of volume-preserving diffeomorphisms of the fluid domain D that map the boundary to itself and go to the identity map at infinity. In (b), $G = \mathrm{SE}(3)$, the finite-dimensional Lie group of rigid body translations and rotations of $\mathbb{R}^3$. The configuration space $Q = G$ in each case, and the phase space is the cotangent bundle of the configuration space, T^*G. In (a) the group action is a right action and in (b) the group action is a left action. In each case, symmetry reduction leads to a Hamiltonian system on the dual of the Lie algebra $\mathfrak{g}^*$ governed by a Poisson bracket termed the Lie–Poisson bracket.[10] The symmetry reduction in (a) is identified with the passage from a Lagrangian or material description of the flow to the Eulerian or spatial description. The symmetry reduction in (b) is identified with the passage from a spatially-fixed frame to a body-fixed frame.

In the dynamically coupled problem of Kirchhoff, both the above Lie groups act but it is difficult to define the configuration space Q as the direct product of these two Lie groups due to the varying fluid domain. One possible way of defining it as the direct product of two groups is as follows. First, note that each $q \in Q$ is a piecewise-smooth embedding of a reference configuration of the body and the fluid into $\mathbb{R}^3$ that maps boundary points to boundary points and goes to the identity map at infinity. Let $\mathrm{Diff}_{\mathrm{vol}}^{\mathrm{SE}(3)}(D_{\mathrm{ref}})$ be the space of volume-preserving diffeomorphisms of D_{ref}, the fluid domain in the reference configuration, that leave the boundary $\partial D_{\mathrm{ref}}$ invariant and with the following important property: every element β of $\mathrm{Diff}_{\mathrm{vol}}^{\mathrm{SE}(3)}(D_{\mathrm{ref}})$ is related to some element $g \in \mathrm{SE}(3)$ by the following condition at infinity:

$$\beta(p) \to \Phi_g(p), \quad p \in D_{\mathrm{ref}}, \; p \to \infty,$$

where $\Phi_g : \mathbb{R}^3 \to \mathbb{R}^3$ denotes the standard rigid body action of $g \in \mathrm{SE}(3)$ on $\mathbb{R}^3$. The element g of $\mathrm{SE}(3)$ is clearly unique, and so accordingly the elements of $\mathrm{Diff}_{\mathrm{vol}}^{\mathrm{SE}(3)}(D_{\mathrm{ref}})$ can be written as $\beta^g \in \mathrm{Diff}_{\mathrm{vol}}^{\mathrm{SE}(3)}(D_{\mathrm{ref}})$ to indicate that the element β^g tends to the map Φ_g at infinity. The map β^g can be extended to a piecewise continuous map on all of $\mathbb{R}^3$ by making it the identity on B if necessary. The configuration space Q can then be identified with a certain infinite-dimensional group $\mathscr{G}$. Define the subgroup $\mathscr{G}$ of the direct product group $\mathrm{SE}(3) \times \mathrm{Diff}_{\mathrm{vol}}^{\mathrm{SE}(3)}(D_{\mathrm{ref}})$ to consist of those pairs of the form $\left(g, \beta^{g^{-1}}\right) \in \mathrm{SE}(3) \times \mathrm{Diff}_{\mathrm{vol}}^{\mathrm{SE}(3)}(D_{\mathrm{ref}})$. Q can

[9] The material on the following two pages is based on notes and discussions between the late Jerrold Eldon Marsden and the author in the period 1998–2000.

[10] The terminology is not due to Arnold but was coined later by Marsden and Weinstein [111]; see [109] for a history of the Lie–Poisson bracket.

then be identified with $\mathscr{G}$ as follows: $q \in Q$ is identified with the pair $\left(g, \beta^{g^{-1}}\right)$, where $g \in \mathrm{SE}(3)$ is uniquely determined by the condition $q\,|_B = \Phi_g \mid B$ and where $\beta^{g^{-1}} = \Phi_{g^{-1}} \circ q$ so that $\beta^{g^{-1}}$ tends to g^{-1} at infinity, as it should. Conversely, given a pair $\left(g, \beta^{g^{-1}}\right) \in \mathrm{SE}(3) \times \mathrm{Diff}_{\mathrm{vol}}^{\mathrm{SE}(3)}(D_{\mathrm{ref}})$, define q to be g on B and set $q = \Phi_g \circ \beta^{g^{-1}}$. Then q tends to the identity at infinity and so $q \in Q$. Group composition in $\mathrm{SE}(3) \times \mathrm{Diff}_{\mathrm{vol}}^{\mathrm{SE}(3)}(D_{\mathrm{ref}})$ is component-wise composition:

$$(h_1, \beta^{g_1}) \cdot (h_2, \beta^{g_2}) = (h_1 \cdot h_2,\ \beta^{g_2} \cdot \beta^{g_1}), \quad \beta^{g_1}, \beta^{g_2} \in \mathrm{Diff}_{\mathrm{vol}}^{\mathrm{SE}(3)}(D_{\mathrm{ref}}), \\ h_1, h_2, g_1, g_2 \in \mathrm{SE}(3).$$

One checks that under this composition law, $\mathrm{SE}(3) \times \mathrm{Diff}_{\mathrm{vol}}^{\mathrm{SE}(3)}(D_{\mathrm{ref}})$, and hence, Q is an infinite-dimensional group. The evolution of the dynamically coupled rigid body-fluid flow system can thus be identified with a time-parametrized curve $q(t) \in Q$ or, equivalently as a pair of curves $(g(t), \beta^{g^{-1}(t)}(t)) \in \mathrm{SE}(3) \times \mathrm{Diff}_{\mathrm{vol}}^{\mathrm{SE}(3)}(D_{\mathrm{ref}})$. It may be noted that the curve $\beta^{g^{-1}}(t)$ corresponds to the flow relative to an observer moving with the body. The above does not constitute a proof that Q is an infinite-dimensional Lie group; for this functional analytic technique such as in, for example, [30, 42] may be required.

An alternative choice of the configuration space, which does not endow it with a group structure, is $Q := \mathrm{Emb}_{\mathrm{vol}}(D_{\mathrm{ref}}, \mathbb{R}^3) \times \mathrm{SE}(3)$. Here, $\mathrm{Emb}_{\mathrm{vol}}(D_{\mathrm{ref}}, \mathbb{R}^3)$ is the space of smooth, volume-preserving embeddings of the domain D_{ref} of a reference configuration of the fluid into $\mathbb{R}^3$, with the usual boundary-to-boundary and identity at infinity conditions. A right action of $\mathrm{Diff}_{\mathrm{vol}}(D_{\mathrm{ref}})$ on Q can then be easily defined in the following way: $(\beta \cdot \varphi, \beta \cdot g) \to (\varphi \circ \beta, g)$, where $\circ$ defines composition of maps, $\varphi \in \mathrm{Emb}_{\mathrm{vol}}(D_{\mathrm{ref}}, \mathbb{R}^3), \beta \in \mathrm{Diff}_{\mathrm{vol}}(D_{\mathrm{ref}})$ and $g \in \mathrm{SE}(3)$. Defining the left action of $\mathrm{SE}(3)$ on Q is a more challenging task due to the infinity condition. The choice of the canonical Φ_g action on $\mathbb{R}^3$, i.e. both body and fluid translated and rotated in a rigid body manner, results in a violation of the infinity condition.

Formally, the symmetry reduced space by the right action of $\mathrm{Diff}_{\mathrm{vol}}(D_{\mathrm{ref}})$ leads to a base space of elements of the form $(\bar{\varphi}, g)$, where $\bar{\varphi}$ may be viewed as the "left-over" part of the map φ after symmetry reduction. It is conjectured that there exists a map—the Kirchhoff flow map—$\alpha : \mathrm{SE}(3) \to \overline{\mathrm{Emb}}_{\mathrm{vol}}(D_{\mathrm{ref}})$, where the latter is the space of maps of the type $\bar{\varphi}$, and that the derivative map $\mathbf{D}\alpha : T\,\mathrm{SE}(3) \to \mathfrak{X}_{\mathrm{K}}$ is given by the linear relation obtained by applying the gradient operator to (1.4); here, $\mathfrak{X}_{\mathrm{K}}$ denotes the space of fluid velocity fields in the spatially-fixed frame in the Kirchhoff problem for a given body shape. This would explain how, by the process of symmetry reduction, Eq. (1.4) allows one to eliminate the fluid terms altogether and capture the entire flow via the added mass tensor with no approximations made. But to the best of the author's knowledge, the existence and properties of the Kirchhoff flow map are still open questions and it is still not completely clear how to understand the passage from Q to Kirchhoff's equations without *a priori* assuming the availability of Eq. (1.4). In other words, it is not clear if and how Eq. (1.4) can be *derived* during a process of symmetry reduction, instead of obtained by inspection as Kirchhoff did. On the other hand, assuming the availability of (1.4), the reduction that takes place is easier to understand and is similar to the reduction in system (b) considered by Arnold. In other words, one starts from the dynamically coupled system in the *spatially-fixed frame* and views this, like the free rigid body, as a system evolving on $T^*\,\mathrm{SE}(3)$. Kirchhoff's equations are then obtained by symmetry reduction of the left action of $\mathrm{SE}(3)$ on this cotangent bundle.

In view of the progress made in recent years in developing Lagrangian and Hamiltonian theories of symmetry reduction in the general, infinite-dimensional setting [32, 106], which lay out in detail all the steps involved in these reduction processes starting from an initial choice of configuration space Q, it is fair to say that there remains much to be done in applying these theories to the infinite-dimensional setting of Kirchhoff's problem and their extensions discussed in the later chapters. The presence of two symmetry groups also suggests the possibility of multiple reduction scenarios. In addition to the issues outlined above, one may also point out that singular vortex models posses degenerate Lagrangians and lack a canonical cotangent bundle Hamiltonian formalism, as discussed in the next three chapters. A clear and detailed application of these theories in the most general setting that sheds light on these issues, among other things, could well be the subject of a whole research monograph in itself.[11]

1.4.1 The Euler–Lagrange and Hamilton's Equations in the Spatially-Fixed Frame

The Lagrangian and Hamiltonian formulations of the dynamically coupled system considered by Kirchhoff in a spatially-fixed frame are not well discussed in the literature. It is therefore instructive to go over these details and see how they lead to Kirchhoff's equations by symmetry reduction of the left action of $\mathrm{SE}(3)$ on $T^*\,\mathrm{SE}(3)$. As mentioned previously, Kirchhoff himself used the least action principle but directly in the body-fixed frame. Modern Lagrangian formulations [32], also applied directly in the body-fixed frame, are discussed in detail in Kanso, Marsden, Rowley and Melli-Huber [79].

First, recall the space of rigid body translations and rotations and their associated linear and angular velocities. Mathematically, this is the space $T\,\mathrm{SE}(3)$, the tangent bundle of $\mathrm{SE}(3)$. The elements of $\mathrm{SE}(3)$ are pairs $(\mathbf{R},\mathbf{x})$, where $\mathbf{R}$ is a rotation matrix and is an element of $\mathrm{SO}(3)$, the space of orthogonal matrices of determinant equal to 1, and $\mathbf{x} \in \mathbb{R}^3$. The Lie algebra of $\mathrm{SO}(3)$ is $\mathfrak{so}(3)$ and is identified with the linear space of skew-symmetric matrices. Its dual space is $\mathfrak{so}(3)^*$, whose elements can also be identified with skew-symmetric matrices. The Lie algebra $\mathfrak{se}(3)$ consists

[11] The currently sparse literature on the subject either focuses on special cases of the problem or/and describes certain basic features incorrectly such as, for example, the left action on $Q := \mathrm{Emb}_{\mathrm{vol}}(D_{\mathrm{ref}},\mathbb{R}^3) \times \mathrm{SE}(3)$, and the identification of the symmetry reduced space by the right action of $\mathrm{Diff}_{\mathrm{vol}}(D_{\mathrm{ref}})$.

of pairs $(\hat{S}, \mathbf{x})$, where $\hat{S} \in \mathfrak{so}(3)$, and its dual $\mathfrak{se}(3)^*$ can also be identified in the same way.

The Overbars Notation

Throughout the book, wherever it is necessary to distinguish variables described in a spatially-fixed frame from variables described in a body-fixed frame, overbars are used to distinguish the former. If this distinction is unnecessary, no overbars are used.

Proposition 1.1. *The Euler–Lagrange equations for the above system, where the Lagrangian $\mathscr{L} : T\,\mathrm{SE}(3) \to \mathbb{R}$ is the system kinetic energy written in a spatially-fixed frame,*

$$\mathscr{L}\left(\mathbf{R}, \bar{\mathbf{r}}_c, \dot{\mathbf{R}}, \bar{\mathbf{V}}\right) = \frac{m}{2}\left\langle \bar{\mathbf{V}}, \bar{\mathbf{V}} \right\rangle_{\mathbb{R}^3} + \frac{1}{2}\left\langle \overline{\mathbf{I}\cdot\Omega}, \bar{\Omega} \right\rangle_{\mathbb{R}^3} - \frac{1}{2}\int_{\partial B} \bar{\Phi}_B \bar{\nabla}\bar{\Phi}_B \cdot \bar{\mathbf{n}}\, v,$$

and $\langle\, , \rangle_{\mathbb{R}^N}$ denotes the standard Euclidean inner product in $\mathbb{R}^N$, are equivalent to Kirchhoff's equations (1.1). In the above $(\mathbf{R}, \bar{\mathbf{r}}_c) \equiv g \in \mathrm{SE}(3)$, $\left(\dot{\mathbf{R}}, \bar{\mathbf{V}}\right) \in T_g\,\mathrm{SE}(3)$.

Proof. The kinetic energy

$$K.E. = \frac{m}{2}\left\langle \bar{\mathbf{V}}, \bar{\mathbf{V}} \right\rangle_{\mathbb{R}^3} + \frac{1}{2}\left\langle \overline{\mathbf{I}\cdot\Omega}, \bar{\Omega} \right\rangle_{\mathbb{R}^3} - \frac{1}{2}\int_{\partial B} \bar{\Phi}_B \bar{\nabla}\bar{\Phi}_B \cdot \bar{\mathbf{n}}\, v$$

is invariant under the transformation from the spatially-fixed to the body-fixed frame, and this is easily checked. Note that $\bar{\Phi}_B$ satisfies the following Neumann problem:

$$\bar{\nabla}^2 \bar{\Phi}_B = 0 \text{ in } D, \quad \bar{\nabla}\bar{\Phi}_B \cdot \bar{\mathbf{n}} = \bar{\mathbf{U}} \cdot \bar{\mathbf{n}} \text{ on } \partial B, \quad \bar{\Phi}_B \to 0 \text{ as } \|\, \bar{\mathbf{r}} \,\| \to \infty, \tag{1.18}$$

where $\bar{\mathbf{U}} = \bar{\mathbf{V}} + \bar{\Omega} \times \bar{\mathbf{l}}$. Position vectors in the two frames are related by

$$\bar{\mathbf{r}} = \mathbf{R}\cdot\mathbf{l} + \bar{\mathbf{r}}_c = \mathbf{R}\cdot(\mathbf{l} + \mathbf{r}_c), \tag{1.19}$$

where $\mathbf{R} \in \mathrm{SO}(3)$ satisfies the orthogonality property, viz. $\left(\mathbf{R}^{-1}\right)^T = \mathrm{Id}$. It follows that

$$\bar{\Phi}_B(\bar{\mathbf{r}}, t) := \Phi_B(\mathbf{l}, t) \equiv \Phi_B\left(\mathbf{R}^{-1}(t)\cdot(\bar{\mathbf{r}} - \bar{\mathbf{r}}_c(t)), t\right), \tag{1.20}$$

since transforming all terms in (1.18) to the body-fixed frame using the above relation gives (1.3). Note that any attempt to linearly decompose $\bar{\Phi}_B$ in terms of corresponding unit potentials in the spatially-fixed frame results in (a) unit potentials that are, in addition, time-dependent and (b) the coefficient of each rigid body velocity component being a linear combination of these unit potentials.

Since $\bar{\mathbf{X}} = \mathbf{R} \cdot \mathbf{X}$, $\bar{\nabla} = \mathbf{R} \cdot \nabla$, and both the volume form and the inner product are invariant under the transformation, the invariance of the system kinetic energy also follows:

$$K.E. = \frac{m}{2} \langle \mathbf{V}, \mathbf{V} \rangle_{\mathbb{R}^3} + \frac{1}{2} \langle \mathbf{I} \cdot \Omega, \Omega \rangle_{\mathbb{R}^3} - \frac{1}{2} \int_{\partial B} \Phi_B \nabla \Phi_B \cdot \mathbf{n} \, \nu,$$

and (1.13) is recovered.

The linear and angular velocities in the two frames are related by

$$\bar{\mathbf{V}} = \dot{\bar{\mathbf{r}}}_c = \mathbf{R} \cdot \mathbf{V}, \quad \dot{\mathbf{R}} = \mathbf{R} \cdot \hat{\Omega} \tag{1.21}$$

with $\hat{\Omega} \in \mathfrak{so}(3)$ the skew-symmetric matrix corresponding to the vector $\Omega \in \mathbb{R}^3$. The Lagrangian of the Proposition can now be written, using (1.12), as

$$\mathscr{L}\left(\mathbf{R}, \bar{\mathbf{r}}_c, \dot{\mathbf{R}}, \bar{\mathbf{V}}\right) = \frac{1}{2} \langle \mathbf{V}, \mathbf{L} \rangle_{\mathbb{R}^3} + \frac{1}{2} \langle \Omega, \mathbf{A} \rangle_{\mathbb{R}^3}, \tag{1.22}$$

and it is clear that this is a function on $T\,\mathrm{SE}(3)$. Recall that for $v \in \mathbb{R}^3$, $\hat{\Omega} \cdot v = \Omega \times v$ and the trace pairing between elements of $\hat{\Omega} \in \mathfrak{so}(3)$ and $\hat{\Omega} \in \mathfrak{so}(3)^*$ is defined as

$$\frac{1}{2} \operatorname{trace}\left(\hat{a}^T \cdot \hat{b}\right) = \langle a, b \rangle_{\mathbb{R}^3}. \tag{1.23}$$

The Lagrangian can therefore also be written as

$$\mathscr{L}\left(\mathbf{R}, \bar{\mathbf{r}}_c, \dot{\mathbf{R}}, \bar{\mathbf{V}}\right) = \frac{1}{2} \langle \mathbf{V}, \mathbf{L} \rangle_{\mathbb{R}^3} + \frac{1}{4} \operatorname{trace}\left(\hat{\Omega}^T \cdot \hat{\mathbf{A}}\right). \tag{1.24}$$

Proceed, in standard fashion, to obtain the Euler–Lagrange equations by implementing the least action principle,

$$\delta \int_{t_1}^{t_2} \mathscr{L}\left(\mathbf{R}, \bar{\mathbf{r}}_c, \dot{\mathbf{R}}, \bar{\mathbf{V}}\right) dt = 0, \tag{1.25}$$

for variations taken over curves $\delta q(t) \equiv (\delta \mathbf{R}(t), \delta \bar{\mathbf{r}}_c(t)) \in T_q Q$ satisfying the fixed end-points condition: $\delta \mathbf{R}(t_1) = \delta \mathbf{R}(t_2) = 0$ and $\delta \bar{\mathbf{r}}_c(t_1) = \delta \bar{\mathbf{r}}_c(t_2) = 0$.

First, obtain the following relations between the variations from (1.21),

$$\begin{aligned} \delta \dot{\bar{\mathbf{r}}}_c &= \delta \mathbf{R} \cdot \mathbf{V} + \mathbf{R} \cdot \delta \mathbf{V}, \\ \delta \dot{\mathbf{R}} &= \delta \mathbf{R} \cdot \hat{\Omega} + \mathbf{R} \cdot \delta \hat{\Omega}. \end{aligned} \tag{1.26}$$

Next, evaluate the left hand side of (6.36) using (1.24):

$$
\begin{aligned}
&\delta \int_{t_1}^{t_2} \mathscr{L}\left(\mathbf{R}, \bar{\mathbf{r}}_c, \dot{\mathbf{R}}, \bar{\mathbf{V}}\right) dt \\
&= \int_{t_1}^{t_2} \left(\langle \delta \mathbf{V}, \mathbf{L} \rangle_{\mathbb{R}^3} + \frac{1}{2} \operatorname{trace}\left(\delta \hat{\Omega}^T \cdot \hat{\mathbf{A}}\right) \right) dt, \\
&= \int_{t_1}^{t_2} \left(\left\langle \left(\mathbf{R}^{-1} \cdot \delta \dot{\bar{\mathbf{r}}}_c - \mathbf{R}^{-1} \cdot \delta \mathbf{R} \cdot \mathbf{V}\right), \mathbf{L} \right\rangle_{\mathbb{R}^3} \right. \\
&\qquad\qquad \left. + \frac{1}{2} \operatorname{trace}\left(\left(\mathbf{R}^{-1} \cdot \delta \dot{\mathbf{R}} - \mathbf{R}^{-1} \cdot \delta \mathbf{R} \cdot \hat{\Omega}\right)^T \cdot \hat{\mathbf{A}} \right) \right) dt.
\end{aligned} \tag{1.27}
$$

Since $\delta \mathbf{R} \in T_g \mathrm{SO}(3)$ $(g \equiv \mathbf{R})$, $\hat{\kappa} := \mathbf{R}^{-1} \cdot \delta \mathbf{R} \in \mathfrak{so}(3)$. Recalling that the Euclidean metric is invariant under the action of the orthogonal matrices $\mathbf{R}$, making the usual assumption of variations commuting with derivatives, implementing the fixed end-points condition and using integration by parts, obtain

$$
\begin{aligned}
&\delta \int_{t_1}^{t_2} \mathscr{L}\left(\mathbf{R}, \bar{\mathbf{r}}_c, \dot{\mathbf{R}}, \bar{\mathbf{V}}\right) dt \\
&= \int_{t_1}^{t_2} \left(\left\langle \delta \bar{\mathbf{r}}_c, -\frac{d(\mathbf{R} \cdot \mathbf{L})}{dt} \right\rangle_{\mathbb{R}^3} - \langle \kappa \times \mathbf{V}, \mathbf{L} \rangle_{\mathbb{R}^3} \right. \\
&\qquad\qquad \left. + \frac{1}{2} \operatorname{trace}\left(\delta \dot{\mathbf{R}}^T \cdot \mathbf{R} \cdot \hat{\mathbf{A}} - \left(\hat{\kappa} \cdot \hat{\Omega}\right)^T \cdot \hat{\mathbf{A}} \right) \right) dt \\
&= \int_{t_1}^{t_2} \left(\left\langle \delta \bar{\mathbf{r}}_c, -\frac{d(\mathbf{R} \cdot \mathbf{L})}{dt} \right\rangle_{\mathbb{R}^3} - \langle \kappa, \mathbf{V} \times \mathbf{L} \rangle_{\mathbb{R}^3} \right. \\
&\qquad\qquad \left. + \frac{1}{2} \operatorname{trace}\left(-\delta \mathbf{R}^T \cdot \frac{d(\mathbf{R} \cdot \hat{\mathbf{A}})}{dt} - \left(\hat{\Omega} \cdot \hat{\kappa} + \widehat{\kappa \times \Omega} \right)^T \cdot \hat{\mathbf{A}} \right) \right) dt.
\end{aligned}
$$

In the last term, use is made of the following relation:

$$
\hat{a} \cdot \hat{b} - \hat{b} \cdot \hat{a} = \widehat{a \times b}, \quad \hat{a}, \hat{b} \in \mathfrak{so}(3).
$$

Continuing,

$$
\begin{aligned}
&\delta \int_{t_1}^{t_2} \mathscr{L}\left(\mathbf{R}, \bar{\mathbf{r}}_c, \dot{\mathbf{R}}, \bar{\mathbf{V}}\right) dt \\
&= \int_{t_1}^{t_2} \left(\left\langle \delta \bar{\mathbf{r}}_c, -\frac{d(\mathbf{R} \cdot \mathbf{L})}{dt} \right\rangle_{\mathbb{R}^3} - \frac{1}{2} \operatorname{trace}\left(\delta \mathbf{R}^T \cdot \mathbf{R} \cdot \widehat{\mathbf{V} \times \mathbf{L}} \right) \right. \\
&\qquad \left. - \frac{1}{2} \operatorname{trace}\left(\delta \mathbf{R}^T \cdot \frac{d(\mathbf{R} \cdot \hat{\mathbf{A}})}{dt} - \delta \mathbf{R}^T \cdot \mathbf{R} \cdot \hat{\Omega} \cdot \hat{\mathbf{A}} + \delta \mathbf{R}^T \cdot \mathbf{R} \cdot \widehat{\Omega \times \mathbf{A}} \right) \right) dt.
\end{aligned} \tag{1.28}
$$

To obtain the last term, one again switches back and forth between the Euclidean and trace pairings.

Collecting terms paired with $\delta\bar{\mathbf{r}}_c$ and $\delta\mathbf{R}^T$, implementing the action principle, and using the last equation in (1.21), one obtains the Euler–Lagrange equations of the system as

$$\frac{d(\mathbf{R}\cdot\mathbf{L})}{dt}=0,$$
$$\mathbf{R}\cdot\frac{d\hat{\mathbf{A}}}{dt}+\mathbf{R}\cdot\widehat{\Omega\times\mathbf{A}}+\mathbf{R}\cdot\widehat{\mathbf{V}\times\mathbf{L}}=0. \tag{1.29}$$

It is obvious that the above equations are equivalent to (1.1). The first equation is the statement of the conservation of the spatial linear momentum of the system in the spatially-fixed frame, $\bar{\mathbf{L}}_S := \mathbf{R}\cdot\mathbf{L}$, and the second equation is the statement of the conservation of the spatial angular momentum of the system, $\bar{\mathbf{A}}_S := \bar{\mathbf{A}}+\bar{\mathbf{r}}_c\times\bar{\mathbf{L}}_S \equiv \mathbf{R}\cdot(\mathbf{A}+\mathbf{r}_c\times\mathbf{L})$.

Hamilton's Equations in the Spatially-Fixed Frame

To write down Hamilton's equations (in the spatially-fixed frame), first obtain the canonical momenta $(\partial\mathscr{L}/\partial\bar{\mathbf{V}},\partial\mathscr{L}/\partial\dot{\mathbf{R}})$. Referring to the derivation of the Euler–Lagrange equations, these are the terms paired with $\delta\dot{\bar{\mathbf{r}}}_c$ and $\delta\dot{\mathbf{R}}$. And so[12]

$$p_L := \frac{\partial\mathscr{L}}{\partial\bar{\mathbf{V}}} = \mathbf{R}\cdot\mathbf{L},$$
$$p_A := \frac{\partial\mathscr{L}}{\partial\dot{\mathbf{R}}} = \mathbf{R}\cdot\hat{\mathbf{A}}.$$

Following convention, the variations of $\dot{\mathbf{r}}_c$ and $\dot{\mathbf{R}}$ are also identified as tangent vectors, and the paired elements as cotangent vectors. The Hamiltonian $\bar{H}: T^*\,\mathrm{SE}(3)\to\mathbb{R}$, where $T^*\,\mathrm{SE}(3)$ is the cotangent bundle of SE(3), is defined by the Legendre transform as

$$\bar{H}\left(\mathbf{R},\bar{\mathbf{r}}_c,p_A,p_L\right)=\langle(p_A,p_L),(\dot{\mathbf{R}},\bar{\mathbf{V}})\rangle_{\mathrm{SE}(3)}-\mathscr{L}\left(\mathbf{R},\bar{\mathbf{r}}_c,\dot{\mathbf{R}},\bar{\mathbf{V}}\right). \tag{1.30}$$

The $\langle\,,\rangle_{\mathrm{SE}(3)}$ pairing between elements $(\alpha,u)\in T_g\,\mathrm{SE}(3)$ and $(\beta,v)\in T_g^*\,\mathrm{SE}(3)$ is defined as follows. Let $\alpha=\mathbf{R}\cdot\hat{a}$ and $\beta=\mathbf{R}\cdot\hat{b}$, where $g\equiv\mathbf{R},\hat{a}\in\mathfrak{so}(3),\hat{b}\in\mathfrak{so}(3)^*$, then

$$\langle(\alpha,u),(\beta,v)\rangle_{\mathrm{SE}(3)} := \frac{1}{2}\operatorname{trace}\left(\hat{a}^T\cdot\hat{b}\right)+\langle u,v\rangle_{\mathbb{R}^3}\equiv\langle a,b\rangle_{\mathbb{R}^3}+\langle u,v\rangle_{\mathbb{R}^3}. \tag{1.31}$$

[12] The reader is cautioned that the symbol p, with and without subscripts, sometimes boldfaced or capitalized, is used extensively throughout the book. But in each instance it should be clear what variable or expression it represents.

As is known [109], Hamilton's equations in canonical form follow naturally from the construction of the Legendre transform. But one may verify this directly "by hand." From (1.30), after expressing $(\dot{\mathbf{R}}, \bar{\mathbf{V}})$ in terms of $(\mathbf{L}, \mathbf{A})$ using (1.12), one immediately obtains

$$\frac{\partial \bar{H}}{\partial p_A} = \mathbf{R} \cdot \hat{\Omega}, \quad \frac{\partial \bar{H}}{\partial p_L} = \mathbf{R} \cdot \mathbf{V}.$$

The partial derivatives of $\bar{H}$ with respect to $\mathbf{R}$ and $\bar{\mathbf{r}}_c$ are the negative of that of $\mathscr{L}$ with respect to these variables. The latter in turn are the terms paired with $\delta\mathbf{R}$ and $\delta\bar{\mathbf{r}}_c$ in the variation of the action integral computed previously (except for those obtained by integration by parts). And so

$$\begin{aligned} \frac{\partial \bar{H}}{\partial \mathbf{R}} &= -\mathbf{R} \cdot \hat{\Omega} \cdot \hat{\mathbf{A}} + \mathbf{R} \cdot \widehat{\Omega \times \mathbf{A}} + \mathbf{R} \cdot \widehat{\mathbf{V} \times \mathbf{L}}, \\ \frac{\partial \bar{H}}{\partial \bar{\mathbf{r}}_c} &= 0 \end{aligned}$$

and Hamilton's equations in the spatially-fixed frame are

$$\begin{aligned} \frac{d\mathbf{R}}{dt} &= \frac{\partial \bar{H}}{\partial p_A}, \\ \frac{d\bar{\mathbf{r}}_c}{dt} &= \frac{\partial \bar{H}}{\partial p_L}, \\ \frac{dp_A}{dt} &= -\frac{\partial \bar{H}}{\partial \mathbf{R}}, \\ \frac{dp_L}{dt} &= -\frac{\partial \bar{H}}{\partial \bar{\mathbf{r}}_c}. \end{aligned} \tag{1.32}$$

The above system is equivalent to Kirchhoff's equations, in addition to having the evolution equations for the body's position and orientation. Formally, this is a Hamiltonian system on $T^*\,\mathrm{SE}(3)$ equipped with the canonical (cotangent) Poisson brackets: for $F, G : T^*\,\mathrm{SE}(3) \to \mathbb{R}$,

$$\{F, G\}_{T^*\,\mathrm{SE}(3)} (\mathbf{R}, \bar{\mathbf{r}}_c, p_A, p_L) := \left(\frac{\partial F}{\partial (\mathbf{R}, \bar{\mathbf{r}}_c)} \frac{\partial G}{\partial (p_A, p_L)} - \frac{\partial F}{\partial (p_A, p_L)} \frac{\partial G}{\partial (\mathbf{R}, \bar{\mathbf{r}}_c)} \right). \tag{1.33}$$

Symmetries, Conserved Quantities, Lie–Poisson Brackets, and More Conserved Quantities

It is well-known that in autonomous Hamiltonian systems, the Hamiltonian function itself is conserved by the dynamics. Other conserved quantities can arise due to the symmetries of the Hamiltonian system. This is of course a consequence of Noether's theorem.

In the dynamically coupled system in the spatially-fixed frame considered above the conservation of $\bar{H}$ is the conservation of the kinetic energy of the rigid body plus fluid. Other conserved quantities arise due to the SE(3) invariance of $\bar{H}$, i.e. due to the invariance under rigid body translations and rotation of the entire fluid+body domain. Formally, this is shown by using the canonical left action L_g of SE(3) on $T^*\,\mathrm{SE}(3)$,

$$L_g : \mathrm{SE}(3) \times T^*\,\mathrm{SE}(3) \to T^*\,\mathrm{SE}(3), \tag{1.34}$$

and showing that $\bar{H}$ is invariant under this action. This is a symmetry of the Hamiltonian system, and Noether's theorem says that the associated momentum map is conserved by the dynamics. Physically, these conserved quantities are identified with the system linear and angular momenta written in a spatially-fixed frame.

$$\begin{aligned} \bar{\mathbf{L}} &:= \mathbf{R}\cdot\mathbf{L} = \mathbf{R}\cdot\left(\mathbf{M}_{11}\ \mathbf{M}_{12}\right)\cdot\left(\mathbf{V}\ \Omega\right)^T, \\ \bar{\mathbf{A}}_s &:= \mathbf{R}\cdot(\mathbf{A}+\mathbf{r}_c\times\mathbf{L}) = \mathbf{R}\cdot\Big(\left(\mathbf{M}_{11}^T\ \mathbf{M}_{22}\right)\cdot\left(\mathbf{V}\ \Omega\right)^T + \mathbf{r}_c\times\mathbf{L}\Big). \end{aligned} \tag{1.35}$$

Apart from giving rise to conserved quantities, symmetries of a Hamiltonian system also have other important consequences. In a famous paper, Marsden and Weinstein [110] showed that in the general setting of a symmetric Hamiltonian system on a symplectic manifold (not necessarily the cotangent bundle of a Lie group) the symmetries can be "eliminated" to obtain a Hamiltonian system on a symmetry reduced manifold that inherits a symplectic form from the original symplectic form. Note that the reduced system still has at least one conserved quantity, the Hamiltonian function restricted to the domain of the reduced manifold. This procedure can also be carried out for symmetric Hamiltonian systems on a Poisson manifold [108]. In this case, the reduced manifold inherits a Poisson bracket from the original one.[13]

In essence, the symmetry reduction techniques split the system's dynamics into two parts—corresponding to motion in the non-symmetry and symmetry directions, respectively. Since typically symmetry directions are known, the former is considered the more important part of the system's dynamics and lies on a manifold that is even "smaller" than the manifold of intersections of the level sets of the conserved quantities. Of course, to obtain the system's total dynamics one has to include the motion in the symmetry directions.

In the particular case when the Poisson manifold is T^*G, the cotangent bundle of a Lie group G, equipped with canonical cotangent Poisson brackets, the symmetry reduced manifold T^*G/G is identified with $\mathfrak{g}^*$, the dual of the Lie algebra of the group, and inherits the Lie–Poisson brackets (as in Arnold's problems). The multiple ways in which the Lie–Poisson brackets are derived from the canonical brackets are explained in [109].

Invoking these theories, Kirchhoff's equations can then be shown to derive from the Hamiltonian system (1.32) governed by (1.33). Since $G = \mathrm{SE}(3)$, the symmetry

[13] A key difference between symplectic reduction and Poisson reduction is that the former also takes into account the conservation of the momentum map.

reduced manifold is $\mathfrak{se}(3)^* \equiv \mathbb{R}^{3*} \times \mathbb{R}^{3*}$ equipped with Lie–Poisson brackets defined as

$$\begin{aligned} \{F, G\}_{\mathfrak{se}(3)^*,\pm}(\mu) &= \pm \left\langle \mu, \left[\frac{\partial F}{\partial \mu}, \frac{\partial G}{\partial \mu} \right] \right\rangle_{\mathfrak{se}(3)}, \\ &= \mp \left\langle \frac{\partial F}{\partial \mu}, \mathrm{ad}^*_{\partial G/\partial \mu} \mu \right\rangle_{\mathfrak{se}(3)}, \end{aligned} \tag{1.36}$$

for $F, G : \mathfrak{se}(3)^* \to \mathbb{R}$ and $\mu \in \mathfrak{se}(3)^*$. The bracket $[\,,\,]$ is the Lie bracket on the Lie algebra $\mathfrak{se}(3)$ and $\langle\,,\,\rangle_{\mathfrak{se}(3)}$ denotes the pairing between $\mathfrak{se}(3)$ and $\mathfrak{se}(3)^*$ given by the right side of (1.31) and which may also be identified with the Euclidean inner product $\langle\,,\,\rangle_{\mathbb{R}^6}$. The ad* operator on $\mathfrak{se}(3)^*$ is given by

$$\mathrm{ad}^*_{(m_1, m_2)}(n_1, n_2) = (n_1 \times m_1 + n_2 \times m_2, n_2 \times m_1). \tag{1.37}$$

With $\mu \equiv (\mathbf{A}, \mathbf{L})$, the Lie–Poisson bracket can be written as

$$\{F, G\}_{\mathfrak{se}(3)^*,\pm}(\mu) = \mp \left\langle \left(\frac{\partial F}{\partial \mathbf{A}}, \frac{\partial F}{\partial \mathbf{L}} \right), \left(\mathbf{A} \times \frac{\partial G}{\partial \mathbf{A}} + \mathbf{L} \times \frac{\partial G}{\partial \mathbf{L}}, \mathbf{L} \times \frac{\partial G}{\partial \mathbf{A}} \right) \right\rangle_{\mathfrak{se}(3)}. \tag{1.38}$$

In Kirchhoff's problem, the Hamiltonian function is the total kinetic energy (1.13) written in terms of $(\mathbf{A}, \mathbf{L})$:

$$H(\mathbf{A}, \mathbf{L}) = \frac{1}{2} (\mathbf{L}, \mathbf{A}) \cdot (\mathbf{M_b} + \mathbf{M_a})^{-1} \cdot (\mathbf{L}, \mathbf{A})^T. \tag{1.39}$$

Since $\partial H / \partial(\mathbf{A}, \mathbf{L}) = (\boldsymbol{\Omega}, \mathbf{V})$, it follows that

$$\mathrm{ad}^*_{\left(\frac{\partial H}{\partial \mathbf{A}}, \frac{\partial H}{\partial \mathbf{L}}\right)}(\mathbf{A}, \mathbf{L}) = (\mathbf{A} \times \boldsymbol{\Omega} + \mathbf{L} \times \mathbf{V}, \mathbf{L} \times \boldsymbol{\Omega}). \tag{1.40}$$

Using the negative Lie–Poisson bracket $\{F, G\}_{\mathfrak{se}(3)^*,-}$ in (1.38), Kirchhoff's equations of motion (1.1) are recovered from the basic definition of a Hamiltonian vector field X_H. Corresponding to a Poisson bracket $\{\,,\,\}$ and a Hamiltonian function H,

$$\dot{F} = \{F, H\}, \tag{1.41}$$

where $\dot{F}$ denotes the rate of change of F along X_H (for the rest of the book, the reference to the sign in the $\mathfrak{se}(3)^*$ Lie–Poisson bracket will be dropped, as it is always the negative bracket that will be used).

In addition to H, Kirchhoff's equations also admit two more conserved quantities. These are Casimir functions, which frequently arise in association with Poisson brackets [109]. Such a function $\mathscr{C}$ Poisson commutes with all functions, i.e. $\{\mathscr{C}, G\} = 0$. A straightforward check shows that the functions given by

$$\mathscr{C}_1 := \frac{1}{2} \langle \mathbf{L}, \mathbf{L} \rangle_{\mathbb{R}^3}, \quad \mathscr{C}_2 := \langle \mathbf{L}, \mathbf{A} \rangle_{\mathbb{R}^3} \tag{1.42}$$

are Casimirs of the Lie–Poisson bracket. Applying (1.38) to $\mathscr{C}_1$, obtain

$$\begin{aligned}\{\mathscr{C}_1, G\} &= \left\langle (0, \mathbf{L}), \mathbf{A} \times \frac{\partial G}{\partial \mathbf{A}} + \mathbf{L} \times \frac{\partial G}{\partial \mathbf{L}}, \mathbf{L} \times \frac{\partial G}{\partial \mathbf{A}} \right\rangle_{\mathfrak{se}(3)}, \\ &= 0\end{aligned}$$

and, similarly,

$$\begin{aligned}\{\mathscr{C}_2, G\} &= \left\langle (\mathbf{L}, \mathbf{A}), (\mathbf{A} \times \frac{\partial G}{\partial \mathbf{A}} + \mathbf{L} \times \frac{\partial G}{\partial \mathbf{L}}, \mathbf{L} \times \frac{\partial G}{\partial \mathbf{A}} \right\rangle_{\mathfrak{se}(3)}, \\ &= 0.\end{aligned}$$

Comparing with the conserved quantities in the spatially-fixed frame (1.35) it may be noted that the conservation of $\mathscr{C}_1$ is implied by the conservation of $\bar{\mathbf{L}}$ due the invariance of the $\mathbb{R}^N$-metric under the action of $\mathbf{R}$.

Perhaps the first paper to recognize the Lie–Poisson structure of Kirchhoff's equations is the one by Novikov and Shmel'tser [122]; see also the paper by Perelomov [128] who attributes it to S. P. Novikov. Using topological and geometric ideas, they investigated the existence of periodic orbits, i.e., periodicity of the orbits $\mathbf{L}(t), \mathbf{A}(t)$. Note that these are not the same as periodic orbits of the rigid body in the physical space. Among other things they characterized the level sets of the above Casimirs as TS^2, the tangent bundle of the two spheres. A discussion of the Lie–Poisson structure of Kirchhoff's equation with the identification of SE(3) as a semi-direct product may be found in Arnold and Khesin [12].

The existence of a fourth conserved quantity, arising under different special cases, was pointed out by Clebsch [35], Steklov, and Lyapunov. These cases are well-summarized in the paper by Rubanovskii [144].

Kozlov and Oniščenko examined the nonintegrability of Kirchhoff's equations in [89]. Aref and Jones [7] presented numerical evidence, using Poincaré sections, of chaotic trajectories. Methods for nonlinear stability analysis of Hamiltonian systems, such as the energy-Casimir and energy-momentum method [9, 12, 105, 109], were applied in [95] and [67]. In the former, the case of non-coincident centers of buoyancy and gravity was studied and in the latter, the case of coincident centers.

The Planar Kirchhoff Problem

In the planar Kirchhoff problem, the rigid body boundary is a closed rigid contour in the plane. Referring to (1.4), there are now only three unit potential functions, ψ_1, ψ_2, and ζ_3, and the added mass tensor of (1.8) reduces to a 3×3 symmetric, positive-definite matrix.

$$\mathbf{M_a} := -\begin{pmatrix} \int_{\partial B} \psi_1 n_1 \ \nu & \int_{\partial B} \psi_2 n_1 \ \nu & \int_{\partial B} \zeta_3 n_1 \ \nu \\ \int_{\partial B} \psi_1 n_2 \ \nu & \int_{\partial B} \psi_2 n_2 \ \nu & \int_{\partial B} \zeta_3 n_2 \ \nu \\ \int_{\partial B} \psi_1 (\mathbf{l} \times \mathbf{n})_3 \ \nu & \int_{\partial B} \psi_2 (\mathbf{l} \times \mathbf{n})_3 \ \nu & \int_{\partial B} \zeta_3 (\mathbf{l} \times \mathbf{n})_3 \ \nu \end{pmatrix}. \tag{1.43}$$

The Lie–Poisson bracket (1.38) continues to apply with $\mathbf{A} \times \partial G/\partial \mathbf{A} = 0$ for all G since both vectors are perpendicular to the plane; in particular, $\mathbf{A}$ and Ω are perpendicular to the plane, so that $\Omega \times \mathbf{A} = 0$. The phase space $(\mathbf{L}, \mathbf{A})$ is three-dimensional and isomorphic to $\mathbb{R}^3$, with the same three invariants, $\mathscr{C}_1, \mathscr{C}_2$ and H. However, $\mathscr{C}_2$ is trivial since $\langle \mathbf{L}, \mathbf{A} \rangle_{\mathbb{R}^3} = 0$ always. The level set of $\mathscr{C}_1$ can be characterized as a cylinder, $S_1 \times \mathbb{R}$, in the $(\mathbf{L}, \mathbf{A})$-space. Since the level set of H is a quadratic surface in the $(\mathbf{L}, \mathbf{A})$-space, the system trajectories will lie on the curves obtained by the intersections of the cylinder and the energy quadratic surface.

In the special case of a circular contour, the rotation of the body has no dynamical effect on the fluid due to the inviscid boundary conditions. In turn, the pressure distribution on the circle can cause no rotation of the body. Therefore, $\Omega(t) = \Omega(0)$ and, without loss of generality, set $\Omega(0) = 0$. This dynamical decoupling also means that $\zeta_3 = 0$, and the circular geometry implies that $\mathbf{l} \times \mathbf{n} = 0$. The potentials ψ_1 and ψ_2 are presented in Eq. (3.43). As a consequence, the added mass matrix simplifies considerably; indeed, it reduces to a diagonal matrix and is identical to the mass matrix

$$\mathbf{M_a} = \mathbf{M_b} = \begin{pmatrix} \pi R^2 & 0 & 0 \\ 0 & \pi R^2 & 0 \\ 0 & 0 & 0 \end{pmatrix}, \tag{1.44}$$

where R is the radius of the circle. $\mathbf{A}$ is no longer a dynamical variable of the system and may be set equal to zero. These configuration simplifications and Kirchhoff's equations together imply that for a circular contour

$$\mathbf{L}(t) = \mathbf{L}(0). \tag{1.45}$$

The body therefore simply translates with its initial velocity and the phase space cylinder reduces to a point. Note that the above equation implies a stronger conservation law than just $\mathscr{C}_1$.

Comparisons may be made with the Blasius theorems of classical hydrodynamics of inviscid, incompressible flows [92, 117]. For a rigid arbitrary shaped (smooth) contour held in place in a steady irrotational streaming flow with velocity $\mathfrak{V}$ at infinity, the hydrodynamic forces on the body are zero in the absence of any circulation around the body. The hydrodynamic moment is zero only for circular contours, but not in general for other shapes. Consistent with this fact, if a non-circular rigid contour is placed in a quiescent fluid and set in motion according to Kirchhoff's equations with $\mathbf{V}(0) = \mathfrak{V}, \Omega(0) = 0$, then generally both $\mathbf{V}$ and Ω will change in time.

1.5 Extending Kirchhoff's Model

In the next few chapters, extensions of Kirchhoff's equations will be considered. These are obtained by relaxing some of the assumptions he made, such as the absence of vorticity, or by adding new features to the domain of the problem, for example, the presence of a dynamic free surface or another rigid body. Viscosity will still be ignored but a later chapter is devoted to describing the experimental and computational work on a dynamically coupled problem involving vortices in a viscous setting. In all chapters, however, it will still be assumed that the fluid is incompressible.

In all these extensions, it will be seen that the generic form of Kirchhoff's equations is still a part of the complete set of equations of the dynamically coupled system with reference to a body-fixed frame. But obviously the system momenta $(\mathbf{L}, \mathbf{A})$ are no longer simply related to the body's velocities via a mass tensor, as in (1.12). Affine terms appear in these equations, involving variables that describe the motion of these additional objects, such as the vortices or the free surface. The complete set of equations has the evolution equations for these objects coupled with the Kirchhoff-like equations. Another common feature, when the models are inviscid, is the presence of Laplace or/and Poisson equations with Dirichlet or Neumann boundary conditions (or both).

From a configurational viewpoint, all the extensions share the following two features: (a) the fluid is external to the body and (b) the body is rigid. But it is important to recognize that Kirchhoff's equations also play an important role in configurations that do not share these features. The most natural one is the complementary problem of a fluid-filled cavity with solid boundaries, in which the fluid is *internal* to the rigid body. This problem too has a history going back to the nineteenth century theoretical mechanics. There is an interesting discussion by Lamb [92] on the "perforated solid" problem (see articles 71 and 72). The incompressible, irrotational flow generated by the rigid motion of the cavity is again governed by Eq. (1.3) but with an important difference: there is no infinity condition to be satisfied. Kirchhoff's decomposition (1.4) still holds, but with the unit potentials $\psi_1 = \psi_2 = \psi_3 = 1$. The lack of the infinity condition means that the (unique) irrotational, incompressible flow generated by the translation component of the body's motion is simply rigid body translation of the fluid. For a partially-filled cavity the fluid surface dynamics becomes the dominant feature—this is the "fluid sloshing" problem that is encountered in a variety of settings, from the familiar sloshing of beverages in bottles to engineering applications such as the sloshing of liquid fuel in rocket fuel tanks; see, for example, [5] for a recent variational approach to a dynamically coupled fluid sloshing problem taking into account vorticity in the fluid. Relaxing feature (b) leads to the dynamically coupled system of a deformable body with an ideal fluid. Such models are more appropriate than single rigid body models for studying problems like the swimming of fish. However, one also needs to come up with an appropriate elastic model for the response of the fish body to the instantaneous fluid stresses. In a series of papers, without assuming any elastic model, Galper and Miloh [47–49] obtain expressions for the fluid dynamic loads on the body due to deformations and in this sense generalize Kirchhoff's equations to deformable bodies, and also discuss Hamiltonian formalism. The general topic of fish swimming is large one, and the reader is referred to the review papers [99, 168] for an overview of modeling efforts, computational and experimental work on this topic and several important references.

1.5.1 The Sum Poisson Bracket

In the inviscid extensions, typically, the Poisson bracket is a sum of the Lie–Poisson bracket on $\mathfrak{se}(3)^*$ and a Poisson bracket that governs the Hamiltonian evolution of these objects with reference to a body-fixed frame. The phase space is a direct product space $P = \mathfrak{se}(3)^* \times A$, where A is the phase space of the interacting objects and also a Poisson manifold. The general theory then guarantees that P is also a Poisson manifold [183].

The sum bracket is defined as follows. Let $\nu \equiv (\mu, \alpha)$. For $F, G : P \to \mathbb{R}$, define functions $F_1, G_1 : \mathfrak{se}(3)^* \to \mathbb{R}$ and $F_2, G_2 : A \to \mathbb{R}$ by projecting on to the respective factors:

$$\begin{aligned} F_1(\mu) &:= F(\mu, \alpha), \quad \alpha \in A \\ F_2(\alpha) &:= F(\mu, \alpha), \quad \mu \in \mathfrak{se}(3)^*, \end{aligned} \tag{1.46}$$

and so on. Implicit and important in these definitions, when applied to a dynamical system, is the assumption that the elements of A and $\mathfrak{se}(3)^*$ can be varied independently of each other.

The sum Poisson bracket is

$$\{F, G\}_P(\nu) = \{F_1, G_1\}_{\mathfrak{se}(3)^*}(\mu) + \{F_2, G_2\}_A(\alpha). \tag{1.47}$$

The four properties that Poisson brackets have to satisfy are bilinearity, skew-symmetry, Leibniz's rule, and Jacobi's identity. The first two are easily seen to be satisfied.

For Leibniz's rule to be satisfied, the following must hold:

$$\{FG, H\}_P(\nu) = F(\nu)\{G, H\}_P(\nu) + \{F, H\}_P(\nu)G(\nu).$$

For the sum bracket, the left hand side is

$$\begin{aligned} \{FG, H\}_P(\nu) &= \{F_1G_1, H_1\}_{\mathfrak{se}(3)^*}(\mu) + \{F_2G_2, H_2\}_A(\alpha), \\ &= F_1(\mu)\{G_1, H_1\}_{\mathfrak{se}(3)^*}(\mu) + \{F_1, H_1\}_{\mathfrak{se}(3)^*} G_1(\mu) \\ &\quad + F_2(\alpha)\{G_2, H_2\}_A(\alpha) + \{F_2, H_2\}_A G_2(\alpha). \end{aligned}$$

Using (1.46) and (1.47), it is easily seen that Leibniz's rule is satisfied.

In Jacobi's identity, it is sufficient to examine a single term:

$$\begin{aligned} &\{\{F, G\}_P, H\}_P(\nu) \\ &= \left\{\{F_1, G_1\}_{\mathfrak{se}(3)^*}, H\right\}_P(\mu) + \{\{F_2, G_2\}_A, H\}_P(\alpha), \\ &= \left\{\left(\{F_1, G_1\}_{\mathfrak{se}(3)^*}\right)_1, H_1\right\}_{\mathfrak{se}(3)^*}(\mu) + \left\{\left(\{F_1, G_1\}_{\mathfrak{se}(3)^*}\right)_2, H_2\right\}_A(\alpha) \\ &+ \{(\{F_2, G_2\}_A)_1, H_1\}_{\mathfrak{se}(3)^*}(\mu) + \{(\{F_2, G_2\}_A)_2, H_2\}_A(\alpha). \end{aligned}$$

But by (1.46), it is obvious that $\left(\{F_1, G_1\}_{\mathfrak{se}(3)^*}\right)_2$ and $(\{F_2, G_2\}_A)_1$ are constant functions and so the corresponding brackets in the above line are zero. Retaining the other two terms, and similarly computing the other terms in Jacobi's identity, it is easily seen that it is satisfied.

An important consequence of having a sum Poisson bracket structure in which one of the brackets is the Lie–Poisson bracket is that the Casimirs $\mathscr{C}_1$ and $\mathscr{C}_2$ continue to be Casimirs of the sum bracket. This results in conserved quantities of the dynamically coupled system in the body-fixed frame.

Chapter 2
The Addition of Vortices

One of the main assumptions made by Kirchhoff in his paper was the absence of vortices [81]:

> Wirbelbewegungen sind in der Flüssigkeit nicht vorhanden.

In the applications discussed in the previous chapter, vortices are somewhat of a common occurrence and are created by the dynamics of the viscous boundary layers on these bodies. Even in the absence of "vortex shedding," there is still vorticity in the boundary layers. However, modeling vorticity effects by directly incorporating viscosity increases the mathematical complexity of the models immensely. Finite-dimensional models are almost impossible without making ad hoc assumptions. Therefore, as a first step to extending Kirchhoff's equations, vorticity is added to the equations but the inviscid framework is retained.

Vorticity fields can be represented in different ways. One could use a continuous vorticity field, with appropriate decay conditions approaching the body and at infinity, or one could use models of vorticity supported on disjoint unions of compact subdomains. Examples of the latter are singular models like point vortices in $\mathbb{R}^2$ and vortex rings in $\mathbb{R}^3$ and non-singular but discontinuous models like vortex patches in $\mathbb{R}^2$ and vortex tubes in $\mathbb{R}^3$. From the point of view of applications, using continuous vorticity fields without prescribing more structure is too general. Patches and tubes have core structures and therefore additional degrees of freedom and, as a consequence, are better models of the vortical structures in real flows that always have core structures. Without gainsaying the importance of representing vorticity by patches and tubes, the book will focus only on extending Kirchhoff's equations using the singular models since they represent the lowest degrees of freedom in $\mathbb{R}^2$ and $\mathbb{R}^3$, respectively, and there are formulational advantages to be gained; for example, it is relatively easier to represent their image fields (see next chapter). Moreover, when it comes to the equations, one expects many of the terms to generalize in expected ways for patches and tubes, for example, sums being replaced by integrals.

This chapter is an introduction to the topic of vorticity and singular vortex models and their Lagrangian and Hamiltonian structure.

B. N. Shashikanth, *Dynamically Coupled Rigid Body-Fluid Flow Systems*,
https://doi.org/10.1007/978-3-030-82646-8_2

2.1 The Importance of Vorticity

Vorticity dynamics constitutes a very important subject in fluid dynamics. Its applications are myriad—extending far beyond well-studied (but not necessarily well-understood) applications in mechanical and aerospace engineering, such as the vortex dynamics in flows over airborne and land vehicles, to applications in superfluids, geophysical and astrophysical fluid dynamics, quantum fluids and relativistic fluid mechanics, to name a few. It is also deemed important for understanding fundamental flow phenomena like turbulence.

Recall that the vorticity field of any flow velocity field $\mathbf{v}$, in the traditional setting of a domain in $\mathbb{R}^3$, is the divergence-free vector field

$$\omega = \nabla \times \mathbf{v}. \tag{2.1}$$

For planar flows, the vorticity vectors are all perpendicular to the plane and the vorticity field can be equivalently identified with a function in the plane. A general and more-encompassing definition views the vorticity as a two-form that is the exterior derivative of the one-form velocity field on a Riemannian manifold M [12]:

$$\omega = \mathbf{d}\mathbf{v}^{\flat}. \tag{2.2}$$

This definition allows the study of vortex dynamics in $\mathbb{R}^n$, $n > 3$, and on non-Euclidean manifolds. Indeed, it is necessary in these higher-dimension spaces to use the 2-form definition, since the identification of vorticity with a field of n-dimensional vectors fails. The divergence-free condition is replaced by the more general condition $\mathbf{dd}\mathbf{v}^{\flat} = 0$, which is a consequence of Poincaré's Lemma. But for most topics in this book, definition (2.1) will suffice.

The evolution of the vorticity field in an inviscid, barotropic flow in $D \subset \mathbb{R}^3$ is governed by the equation:

$$\frac{\partial \omega}{\partial t} + \mathbf{v} \cdot \nabla \omega = \omega \cdot \nabla \mathbf{v}, \tag{2.3}$$

coupled with initial conditions and (if necessary) boundary conditions. This equation clearly shows that the evolution is influenced by the instantaneous velocity field associated with the vorticity field. In other words, it becomes important to invert equation (2.1). This leads to the famous *Biot–Savart law* for vorticity fields of incompressible flows:

$$\mathbf{v}(p) = -\int_D (\nabla \times \omega)(\tilde{p}) G(p, \tilde{p})\, \tilde{\mu}, \quad p, \tilde{p} \in D, \tag{2.4}$$

where G is the Green's function of the domain D.

The importance of vorticity in dynamically coupled problems may be easily seen from the following vector identities.[1] For a vector field $\mathbf{v}$ in a bounded domain $D \subset \mathbb{R}^3$, keeping in mind the convention for unit normals introduced in the previous chapter,

$$\int_D \mathbf{v}\,\mu = \frac{1}{2}\int_D \mathbf{r}\times\omega\,\mu + \frac{1}{2}\int_{\partial D} \mathbf{r}\times(\mathbf{n}\times\mathbf{v})\,\nu. \tag{2.5}$$

For a fluid of homogeneous density ρ and velocity field $\mathbf{v}$, the integral on the left is the fluid linear momentum (modulo the constant density). Clearly in the interior of the domain the contribution to the fluid linear momentum is solely due to the vorticity. Note that this conclusion requires only that the density to be constant and is equally valid for inviscid and viscous flows. But it should be remembered that in a viscous flow, the viscosity influences the evolution of the vorticity. If one applies this identity to a viscous flow with rigid boundaries satisfying a no-slip boundary condition, then the rigid boundary's velocity field directly enters this equation via the boundary integral. This is discussed more in Chap. 5.

An analogous identity,

$$\int_D \mathbf{r}\times\mathbf{v}\,\mu = -\frac{1}{2}\int_D r^2\omega\,\mu - \frac{1}{2}\int_{\partial D} r^2(\mathbf{n}\times\mathbf{v})\,\nu, \tag{2.6}$$

and another identity also obtained by integration by parts,

$$\int_D \langle\mathbf{v},\mathbf{v}\rangle_{\mathbb{R}^3}\,\mu = \int_D \langle\mathbf{A},\omega\rangle_{\mathbb{R}^3}\,\mu - \int_{\partial D} \mathbf{A}\times\mathbf{v}\cdot\mathbf{n}\,\nu, \tag{2.7}$$

where $\mathbf{A}$ is a divergence-free vector potential[2] satisfying

$$\mathbf{v} = \nabla\times\mathbf{A}, \tag{2.8}$$

$$\nabla^2\mathbf{A} = -\omega \tag{2.9}$$

show that the same comments about the vorticity apply to the fluid's angular momentum and kinetic energy as well. The domain integrals involving vorticity in (2.5) and (2.6) are termed the first and second moments of vorticity, respectively, and the contour integrals are termed the first and second moments of circulation, respectively. It may be noted, after using some standard vector calculus, that a change by $\mathbf{b}\in\mathbb{R}^3$ in the origin of the reference frame changes terms consistently. The linear momentum is invariant and the angular momentum changes by $\mathbf{b}\times\int_D \omega\,\mu$. These two equations are valid even when $\nabla\cdot\mathbf{v}\neq 0$. However, when the varying density field of compressible flows is included in the integrals on the left hand sides, there are additional interior terms on the right hand sides that contribute to the vorticity

[1] These are easily derived from the classical vector integral theorems of calculus but can also be found, for example, in Saffman's book [147].

[2] The symbol used here is identical to that used for the angular momentum. However, in the rest of the book, the vector potential symbol will always have subscripts or superscripts making it easy to distinguish it from the angular momentum.

and that involve the gradients of the density field. Finally, a word of caution is in order when applying (2.5) to planar domains: the factor $1/2$ is absent.

These identities clearly show that the distribution of vorticity in the fluid domain interior plays a key role in the way in which the body harnesses momentum and kinetic energy from the fluid.

2.2 Singular Vortex Models

Since Helmholtz formulated his famous laws of inviscid barotropic vortex motion [59, 60], it has been well-known that vorticity dynamics in such a setting has interesting geometric features not present in the corresponding velocity fields, such as, for example, the persistence of vortex tubes and their linking in $\mathbb{R}^3$. Mathematically too, there are some known advantages of studying vorticity evolution equations in this setting; for example, the pressure term is absent and Kelvin's circulation theorem is valid. Arnold [8, 10] and later, Marsden and Weinstein [111], among others, showed how to place the vorticity evolution equations (2.3) in a modern Hamiltonian framework. Indeed, these equations are the Lie–Poisson equations of Chap. 1 when $G = \mathrm{Diff}_{\mathrm{vol}}(D)$.

The dynamics and geometry of singular vortex models, in particular, have attracted the attention of mathematicians, physicists, and engineers alike. In such models, infinite vorticity is concentrated on discrete subdomains and the mutual and self-interaction dynamics of these vortices is analyzed. Specifically, the vorticity field is modeled as a sum of Dirac delta distributions supported on codimension-2 surfaces. In $\mathbb{R}^2$, for example, these surfaces are points and are commonly known as *point vortices*:

$$\omega(\mathbf{r};\mathbf{r}_j) = \sum \Gamma_j \mathbf{k} \delta(\mathbf{r} - \mathbf{r}_j), \tag{2.10}$$

where $\mathbf{r}_j = (x_j, y_j)$ is the location of the jth point vortex in the domain and $\mathbf{k}$ is the unit vector perpendicular to $\mathbb{R}^2$.[3] In $\mathbb{R}^3$, for example, these surfaces are smooth curves and are commonly known as *vortex filaments*:

$$\omega(\mathbf{r};\mathbf{r}_j(s_j)) = \sum \Gamma_j \mathbf{t}_j(s_j) \delta(\mathbf{r} - \mathbf{r}_j(s_j)), \tag{2.11}$$

with

$$\mathbf{r}_j(s_j) \equiv C_j(s_j). \tag{2.12}$$

In the above, $C_j : S^1\,(\text{or } \mathbb{R}) \times \mathbb{R} \to \mathbb{R}^3$ is the map that defines the parametrized jth filament, $\mathbf{t}_j$ is its unit tangent vector field, and $\mathbf{r}_j(s_j) = (x_j(s_j), y_j(s_j), z_j(s_j))$ denotes the coordinates of points on the filament. In both the above equations, Γ_j stands for

[3] To avoid notational clutter, the index will not be indicated in single summations. Neither will the range, unless it is different from 1 to N.

the strength of the jth vortex and is equal to the circulation $\oint_c \mathbf{v} \cdot d\mathbf{r}$ about any closed loop c enclosing the vortex. As a consequence of the Helmholtz–Kelvin laws of vortex motion, Γ_j is a Lagrangian invariant of the flow. The identification in (2.11) will be invoked occasionally to switch between the notations $\mathbf{r}_j$ and C_j (or between $\mathbf{l}_j$ and C_j in the case of a body-fixed frame).

Notation

In the above and in what follows, the time-dependency of $\mathbf{r}_j, C_j$ (and other time-dependent curve variables) is always implied and extra symbols to denote this will be avoided, i.e. $\mathbf{r}_j(s_j), C_j(s_j)$ etc. will be used instead of the more precise notation $\mathbf{r}_j(s_j,t), C_j(s_j,t)$ etc. Moreover, the symbol C_j will be sometimes used for the image of the map as well; for example, $p \in C_j$ will mean point p lying on the jth filament.

2.2.1 The *N*-Point-Vortex Model

The dynamics of point vortices in the unbounded plane, or the N-point-vortex model, has been quite well studied for $N \leq 4$. This model again goes back to the days of Kelvin, Helmholtz and Kirchhoff. Indeed it was Kirchhoff again who made a pioneering contribution, showing that the equations of motion of N-point-vortices in the plane can be written as a system of $2N$ nonlinear ordinary differential equations [82]:

$$\Gamma_j \frac{d\mathbf{r}_j}{dt} = -J\frac{\partial H}{\partial \mathbf{r}_j}, \quad j = 1, \cdots, N, \tag{2.13}$$

where J is the matrix:

$$J = \begin{pmatrix} 0 & -1 \\ 1 & 0 \end{pmatrix}, \tag{2.14}$$

and

$$H(\mathbf{r}_j) = -\frac{1}{4\pi} \sum_{i,j(j>i)} \Gamma_i \Gamma_j \log\left[(x_i - x_j)^2 + (y_i - y_j)^2\right]. \tag{2.15}$$

It may be shown—see, for example, Batchelor [15]—that the above function is the kinetic energy of the flow generated by the vortices after one subtracts two infinity terms: (a) the unbounded growth of the kinetic energy associated with the velocity field of each vortex as one approaches that vortex or, in short, the infinite self-interaction kinetic energy, and (b) the unbounded growth of the kinetic energy as one approaches infinity in the physical domain. The latter infinity occurs, however, only if the sum of the point vortex strengths is different from zero. More details of this identification are presented in Sect. 3.1.

Kirchhoff also showed that, apart from H, the following functions are also conserved by the motion:

$$L_x = \sum \Gamma_j y_j, \quad L_y = \sum \Gamma_j x_j, \quad A = -\frac{1}{2}\sum \Gamma_j \left(x_j^2 + y_j^2\right). \tag{2.16}$$

In modern parlance, the N-point-vortex model is a Hamiltonian system on the phase space $P_{\mathrm{pv}} = \mathbb{R}^{2N} \backslash \triangle$, where the excluded set $\triangle$ is the set of all collision points of the vortices, with the Hamiltonian function $H : P_{\mathrm{pv}} \to \mathbb{R}$ given by (2.15) and equipped with the symplectic form $\Omega_{\mathrm{pv}}(p) : T_p P_{\mathrm{pv}} \times T_p P_{\mathrm{pv}} \to \mathbb{R}$ ($T_p P_{\mathrm{pv}} \equiv \mathbb{R}^{2N}$) given by

$$\Omega_{\mathrm{pv}}(p)(u,v) := \sum \Gamma_j dx_j \wedge dy_j(u,v), \quad p \in P_{\mathrm{pv}}, \quad u,v \in T_p P_{\mathrm{pv}} \tag{2.17}$$

(and whose symplectic matrix is $-J$.) The Poisson bracket corresponding to this symplectic form is

$$\{F,G\}_{\mathrm{pv}}(p) := \sum \frac{1}{\Gamma_j}\left(\frac{\partial F}{\partial x_j}\frac{\partial G}{\partial y_j} - \frac{\partial F}{\partial y_j}\frac{\partial G}{\partial x_j}\right), \quad p \in P_{\mathrm{pv}}, \quad F,G : P_{\mathrm{pv}} \to \mathbb{R}. \tag{2.18}$$

The three conserved quantities in (2.16) are the components of the momentum map $\mathbf{J} : P_{\mathrm{pv}} \to \mathfrak{se}(2)^*$ of the SE(2)action on P_{pv}, arising due to invariance of H under this action and is a consequence of Noether's Theorem. These quantities are related to the linear momentum and angular momentum of the flow generated by the vortices, and these relations are discussed more in Sect. 3.2.2.

There also exists a fifth conserved quantity, see Lamb [92] (Art. 157),

$$V(\mathbf{r}_1, \cdots, \mathbf{r}_N, \dot{\mathbf{r}}_1, \cdots, \dot{\mathbf{r}}_N) := \sum \Gamma_j \left(x_j \frac{dy_j}{dt} - y_j \frac{dx_j}{dt}\right), \tag{2.19}$$

related to scaling invariance. This invariance may be understood in terms of Euler's famous result on homogeneous functions.[4] First, note that (2.15) can be written as

$$-2\pi H = \log\left[\prod_{\substack{i,j \\ (j>i)}} r_{ij}^{\Gamma_i \Gamma_j}\right], \quad r_{ij} = | \mathbf{r}_i - \mathbf{r}_j |.$$

It follows that

$$h(\mathbf{r}_1, \cdots, \mathbf{r}_N) := e^{-2\pi H(\mathbf{r}_1, \cdots, \mathbf{r}_N)}$$

4 For an interesting dynamics feature related to symmetries and this homogeneous function property of the point vortex Hamiltonian, see [62].

is an homogeneous function of $(x_1, y_1, \cdots, x_N, y_N)$ of degree $\sum\limits_{\substack{i,j \\ (j>i)}} \Gamma_i \Gamma_j$. Clearly, the conservation of H implies the conservation of h and vice versa. Applying Euler's result,

$$\sum \mathbf{r}_j \cdot \frac{\partial h}{\partial \mathbf{r}_j} = \Big(\sum_{\substack{i,j \\ (j>i)}} \Gamma_i \Gamma_j \Big) h,$$

$$\Rightarrow -2\pi e^{-2\pi H} \sum \mathbf{r}_j \cdot \frac{\partial H}{\partial \mathbf{r}_j} = \Big(\sum_{\substack{i,j \\ (j>i)}} \Gamma_i \Gamma_j \Big) h$$

shows that the sum on the left is conserved by the dynamics, and the conservation of V follows from (2.13).

Euler–Lagrange Equations for the N-Point-Vortex Model

Chapman [34] used V to define a conserved Lagrangian, $\mathscr{L} : TP_{\mathrm{pv}} \to \mathbb{R}$, for the N-point-vortex model as

$$\mathscr{L}(\mathbf{r}_1, \cdots, \mathbf{r}_N, \dot{\mathbf{r}}_1, \cdots, \dot{\mathbf{r}}_N) := -\frac{1}{2} V(\mathbf{r}_1, \cdots, \mathbf{r}_N, \dot{\mathbf{r}}_1, \cdots, \dot{\mathbf{r}}_N) - H(\mathbf{r}_1, \cdots, \mathbf{r}_N). \quad (2.20)$$

The corresponding Euler–Lagrange equations are

$$\frac{d}{dt}\left(\frac{\partial \mathscr{L}}{\partial \dot{\mathbf{r}}_j}\right) - \frac{\partial \mathscr{L}}{\partial \mathbf{r}_j} = 0, \quad j = 1, \cdots, N,$$

$$\Rightarrow \left(\frac{1}{2}\Gamma_j \frac{dy_j}{dt}, -\frac{1}{2}\Gamma_j \frac{dx_j}{dt}\right) - \left(-\frac{1}{2}\Gamma_j \frac{dy_j}{dt} - \frac{\partial H}{\partial x_j}, \frac{1}{2}\Gamma_j \frac{dx_j}{dt} - \frac{\partial H}{\partial y_j}\right) = 0, \quad j = 1, \cdots, N,$$

which are identical to (2.13).

However, Chapman's Lagrangian, being linear in the point vortex velocities, is *degenerate* [143]. The canonical momenta are obtained as

$$\mathbf{p}_j := \partial \mathscr{L} / \partial \dot{\mathbf{r}}_j = (\Gamma_j y_j, -\Gamma_j x_j)/2 \quad (2.21)$$

and are formally identified with cotangent vectors, i.e. as elements lying in the cotangent space. Since $P_{\mathrm{pv}} \subset \mathbb{R}^{2N}$, these variables can be identified with the following linear map, $\mathbf{p} : T_p P_{\mathrm{pv}} \to \mathbb{R}$, using the standard inner product $< , >_{\mathbb{R}^{2N}}$:

$$\mathbf{p}(\mathbf{u}) := < \mathbf{p}_j, \mathbf{u}_j >_{\mathbb{R}^{2N}} = \left(\sum \Gamma_j y_j u_{j,x} - \sum \Gamma_j x_j u_{j,y} \right)/2, \quad (2.22)$$

where $\mathbf{u} \equiv (\mathbf{u}_1, \cdots, \mathbf{u}_N) \in T_p P_{\mathrm{pv}}, \mathbf{u}_j \equiv (u_{j,x}, u_{j,y})$. This then allows $\mathbf{p} \equiv (\mathbf{p}_1, \cdots, \mathbf{p}_N)$ to be identified as an element of the cotangent space $T_p^* P_{\mathrm{pv}}$. However, the expression (2.21) for the canonical momenta is rather unusual, since the momenta are independent of the velocities and depend on the base point p. Recall that the canonical momenta are actually fiber derivatives [109], which means that at each $p \in P_{\mathrm{pv}}$ they define a map between $T_p P_{\mathrm{pv}}$ and $T_p^* P_{\mathrm{pv}}$:

$$(\dot{\mathbf{r}}_1, \cdots, \dot{\mathbf{r}}_N)(p) \mapsto (\mathbf{p}_1, \cdots, \mathbf{p}_N)(p).$$

It follows that (a) the above is a constant map, and hence not invertible, and (b) $(\mathbf{p}_1, \cdots, \mathbf{p}_N)$ and $(\mathbf{r}_1, \cdots, \mathbf{r}_N)$ are not independent. Legendre's transform therefore does not produce equations of motion in a canonical Hamiltonian form on the cotangent bundle.

One can also understand this from a more geometric mechanics viewpoint. Equation (2.21) or (2.22) may be viewed as defining the section of a cotangent bundle, $\mathscr{S} : P_{\mathrm{pv}} \to T^* P_{\mathrm{pv}}$. Let $\mathscr{R} := \mathrm{Range}(\mathscr{S})$. Clearly, this is a space isomorphic to P_{pv}. Now $P_{\mathrm{pv}} \times \mathscr{R} \subset T^* P_{\mathrm{pv}}$ and let $i : P_{\mathrm{pv}} \times \mathscr{R} \to T^* P_{\mathrm{pv}}$ denote the inclusion. If Ω_{can} denotes the canonical symplectic form on $T^* P_{\mathrm{pv}}$, then it is easily checked that $i^* \Omega_{can}$—the two-form on $P_{\mathrm{pv}} \times \mathscr{R}$ obtained by pullback—is not symplectic, since it is degenerate. However, the two-form obtained as $\mathscr{S}^* i^* \Omega_{can}$ is symplectic and is equal to (2.17), i.e.

$$\Omega_{\mathrm{pv}} = \mathscr{S}^* i^* \Omega_{can}. \tag{2.23}$$

With the development of modern theories of dynamical systems, there was revival of interest in this model, especially with the papers of Ziglin [194] and Koiller and Carvalho [86], who provided analytic proofs that the motion of N vortices, when N=4, could be non-integrable. Prior to this Novikov and Sedov [121] presented an example of three point vortices collapsing; see also [61] and [90] for more recent investigations of collapse involving more than three vortices in unbounded domains, and Flucher and Gustafsson [45] for collapse and collision of vortices in bounded domains. Given the importance of vorticity in turbulence, interest arose in the research community whether this model could provide more insights into the phenomenon of turbulence and into the related issue of integrability of Euler's equations for a fluid flow. These weighty issues aside though the model has remained of interest to mathematicians, physicists, and engineers. The reasons are not hard to understand. Fluid flows governed by Euler's equations are generically infinite-dimensional systems, and the N-point-vortex model is a rare example of a finite-dimensional representation making the equations more amenable to the tools of dynamical systems and also easier to numerically integrate. And despite the strong singularities in the model, the motion of the vortices itself gives rise to non-singular dynamics (except in some non-generic cases, as mentioned above). For compilations of important results and comprehensive bibliographies on the subject, see [112, 120, 160].

2.2.2 The N Vortex Ring Model

Real flows occur in three-dimensional space. As is well-known, vorticity fields in such flows have some important features that are not captured by planar vorticity fields [126].

The notion of point vortices does not carry over to $\mathbb{R}^3$. It is clear that a field of vorticity vectors of any magnitude (finite or infinite) based at isolated points in $\mathbb{R}^3$ fails to satisfy the divergence-free condition. There is also no canonical way of prescribing the direction of these vorticity vectors. Equivalently, the corresponding vorticity two-form fails to satisfy Poincaré's Lemma and there is no canonical way of prescribing a local oriented plane on which the two-form acts.

Vortex filaments are a natural way of extending the notion of point vortices. As briefly described in Sect. 2.2, the vorticity 2-form is now a Dirac delta distribution supported on curves in $\mathbb{R}^3$ (assumed to be smooth and non-intersecting). Indeed, a point vortex may be thought of as the projection on a perpendicular plane of an infinitely long, straight vortex filament in $\mathbb{R}^3$. A vortex ring refers to a filament that is closed. It is a consequence of the divergence-free nature of vorticity that a filament that is not closed or does not end on a boundary must be infinitely long; see Velasco-Fuentes [175] for an interesting discussion on this topic.

The vorticity distribution represented by filaments has a three-dimensional structure. Though still missing core structure and the associated dynamics, filaments do exhibit some important features missing in point vortex models, namely, curvature of the filaments and the self-induced velocity field. They are more appropriate models for the tube-like vortex structures that occur in some real flows such as, for example, tornadoes, trailing wing-tip vortices of aircraft, smoke rings, and water rings.

Self-Induced Velocity Field and Its Regularization

Before discussing the Hamiltonian model of N rings, it is important to recall an important feature of filaments missing in point vortices—*the self-induced velocity field*, due to the curvature of filaments. Applying (2.4), using the Green's function in $\mathbb{R}^3$, gives the velocity field due to a filament C as

$$\mathbf{v}(p) = \frac{\Gamma}{4\pi}\int_C \frac{\mathbf{t}(s)\times(\mathbf{r}(p)-\mathbf{r}(s))}{|\mathbf{r}(p)-\mathbf{r}(s)|^3}ds, \quad p\in\mathbb{R}^3. \tag{2.24}$$

For a point that lies on the filament, $p_\mathrm{f}\in C$ with parameter value denoted by s_f, the expression shows that the cancellation of induced velocity that occurs for a straight filament cannot be expected to occur, in general, for a curved filament. And so one expects a non-zero induced velocity at any point on the filament. However, the integral has a logarithmic singularity and gives an infinite value of the induced velocity at each point. Indeed, assuming the terms $\mathbf{t}(s)$ and $\mathbf{r}(s)$ are infinitely differentiable for all s, expanding in powers of $s-s_\mathrm{f}$ gives

$$
\begin{aligned}
\mathbf{v}_{\mathrm{SI}}(p_{\mathrm{f}}) &= \frac{\Gamma}{4\pi}\int_C \left(\frac{\left(\mathbf{t}\times\frac{d\mathbf{t}}{ds}\right)(s_{\mathrm{f}})}{2\mid s-s_{\mathrm{f}}\mid} + O(1) \right) d(s-s_{\mathrm{f}}), \quad p_{\mathrm{f}}\in C, \\
&= \frac{\Gamma}{4\pi}(\kappa\mathbf{b})(s_{\mathrm{f}})\int_C \frac{d(s-s_{\mathrm{f}})}{2\mid s-s_{\mathrm{f}}\mid} + \frac{\Gamma}{4\pi}\int_C O(1)\, d(s-s_{\mathrm{f}}), \quad p_{\mathrm{f}}\in C, \qquad (2.25)
\end{aligned}
$$

assuming s to be the curve arc-length parameter and where the Serret–Frenet equations for a curve have been used; κ denotes the principal curvature and $\mathbf{b}$ the binormal direction. The leading integral has logarithmic divergence as $s \to s_{\mathrm{f}}$ and therefore to obtain meaningful estimates of $\mathbf{v}_{\mathrm{SI}}(p_{\mathrm{f}})$ requires regularization techniques.

One of the more commonly used methods of regularization of the above integral is the *local induction approximation (LIA)*. This method goes back to DaRios, and a very informative history of this method and its independent re-discovery by others may be found in the articles by Ricca [134, 135]. A brief description of the LIA is given here but the reader is encouraged to look up these original articles for a fuller understanding. The LIA takes its name by assuming that $\mathbf{v}_{\mathrm{SI}}$ is determined primarily by portions of the filament adjacent to p_{f} (on both sides). $O(1)$ contributions that come from portions of the filament farther away are neglected. The divergent contribution of the leading order integral is then regularized by introducing a small cut-off parameter c as follows:

$$
\begin{aligned}
\int_C \frac{d(s-s_{\mathrm{f}})}{2\mid s-s_{\mathrm{f}}\mid} &\approx \int_{-a}^{-c} \frac{d(s-s_{\mathrm{f}})}{2\mid s-s_{\mathrm{f}}\mid} + \int_c^a \frac{d(s-s_{\mathrm{f}})}{2\mid s-s_{\mathrm{f}}\mid} \\
&= \frac{1}{2}\left(-\log\mid s-s_{\mathrm{f}}\mid_{-a}^{-c} + \log\mid s-s_{\mathrm{f}}\mid_c^a\right) \\
&\approx -\log\mid c\mid .
\end{aligned}
$$

Under the LIA, the self-induced velocity is therefore written as

$$
\mathbf{v}_{\mathrm{SI,reg}}(p_{\mathrm{f}}) = \mathscr{K}\,\kappa\mathbf{b}(s_{\mathrm{f}}), \qquad (2.26)
$$

where $\mathscr{K}$ is related to the strength of the filament and the cut-off parameter and is typically assumed to be constant. The determination of the value of the cut-off parameter remains somewhat arbitrary. The most realistic estimates are those that are obtained by applying the Biot–Savart law (2.4) to a vortex tube with a thin core and using asymptotic techniques to evaluate the leading order term for the self-induced velocity field as the core size goes to zero [29, 147].

Despite this arbitrariness in the choice of the cut-off parameter, Eq. (2.26) has attracted a lot of attention among physicists and mathematicians due to the presence of the curve-intrinsic parameters κ and $\mathbf{b}$. Written as an evolution equation for the filament, with $\mathscr{K}$ assigned the value of unity,

$$
\frac{\partial C}{\partial t} = \kappa\mathbf{b}, \qquad (2.27)
$$

it becomes an intrinsic evolution equation for a geometric curve in $\mathbb{R}^3$ and can then be analyzed regardless of its origins in vortex dynamics. It is worth pointing out that these curve-intrinsic parameters already appear in (2.25) even before the regularization. Equation (2.27) attracted even more interest after Hasimoto [58] showed that the equation can be transformed into the 1D nonlinear Schrödinger equation.

The binormal singularity is also obtained if one starts with the expression for the velocity field due to an isolated curved filament (2.24) and considers the limiting process of a field point approaching the filament, i.e. $p \to p_\mathrm{f}$, as discussed in Appendix B. This process also reveals, in addition, a point vortex type singularity. But just as for point vortices this singularity results in zero self-induced velocity for the filament. Both these singularities result in making the kinetic energy of the flow due to an isolated filament infinite:

$$K.E._{\mathrm{SI}} := \frac{1}{2}\int_{\mathbb{R}^3} \langle \mathbf{v}_\mathrm{f}(p), \mathbf{v}_\mathrm{f}(p)\rangle_{\mathbb{R}^3}\, \mu = \infty.$$

Subtracting the kinetic energy of the point vortex term and regularizing the binormal term lead to a finite self-induced kinetic energy $K.E._{\mathrm{SI,reg}}$.

The presence of a non-zero self-induced velocity field for a curved vortex filament implies that an isolated curved vortex filament moves, unlike an isolated point vortex or an isolated straight line vortex. And this self-interaction should also be included in a model of mutually interacting N filaments.

Hamiltonian Model of N Rings

Marsden and Weinstein [111] were the first to present the symplectic structure of filament motion. Some basic constructs involved in the Hamiltonian formulation of a filament may be recalled. The phase space of a filament, P_f, is infinite-dimensional and is identified with the space of images of parametrized curves $C : \mathbb{R}(\text{or } S^1) \times \mathbb{R} \to \mathbb{R}^3$. This implies that two different parametrizations of the curve with the same image in $\mathbb{R}^3$ correspond to the same point $p \in P_\mathrm{f}$. A tangent vector $u \in T_pP_\mathrm{f}$ is therefore identified with a vector field in $\mathbb{R}^3$, based on the curve, whose elements are normal to the curve at each point (possibly zero at some points). The zero element in T_pP_f is the equivalence class of all vector fields, based on the curve, which are tangent to the curve everywhere.

The Marsden–Weinstein symplectic form $\Omega_\mathrm{f}(p) : T_pP_\mathrm{f} \times T_pP_\mathrm{f} \to \mathbb{R}$, for a ring of strength Γ, is

$$\Omega_\mathrm{f}(p)(u,v) = \Gamma \oint_C \langle \mathbf{t}(s), u(s) \times v(s)\rangle_{\mathbb{R}^3}\, ds, \quad p \in P_\mathrm{f}, \quad u, v \in T_pP_\mathrm{f}.$$

The associated canonical Poisson bracket is written as

$$\{F,G\}_\mathrm{f} = \frac{1}{\Gamma}\oint_C \left\langle \mathbf{t}, \left(\frac{\delta F}{\delta C}\right)^\sharp \times \left(\frac{\delta G}{\delta C}\right)^\sharp \right\rangle_{\mathbb{R}^3} ds, \quad \frac{\delta F}{\delta C}, \frac{\delta G}{\delta C} \in T_p^*P_\mathrm{f}, \tag{2.28}$$

where $T_p^* P_\mathrm{f}$ denotes the cotangent space at p and the functional derivatives are identified with 1-forms in $\mathbb{R}^3$ based on the ring. The $\sharp$ superscript denotes the associated vector fields, based on the ring, using the canonical metric in $\mathbb{R}^3$. To avoid more notation, the same symbols are used for the elements of $T_p P_\mathrm{f}$ and the corresponding vector fields in $\mathbb{R}^3$: similarly for the elements of $T_p^* P_\mathrm{f}$ and the corresponding 1-forms in $\mathbb{R}^3$.

To understand these notions better, first note that a variational vector field δC is identified with a *normal* vector field based on the ring. The functional derivatives are then defined according to the formula:

$$\lim_{\varepsilon\to 0}\frac{1}{\varepsilon}\left(F(p+\varepsilon p)-F(p)\right)=\oint_C \frac{\delta F}{\delta C}(\delta C)\,ds \equiv \oint_C \left\langle \left(\frac{\delta F}{\delta C}\right)^\sharp, \delta C\right\rangle_{\mathbb{R}^3} ds. \tag{2.29}$$

The last two terms are also, equivalently, the definition of the pairing $\langle\,,\rangle_{P_\mathrm{f}}$ between the elements of $T_p P_\mathrm{f}$ and $T_p^* P_\mathrm{f}$. Note that an alternative way of writing (2.28) is

$$\{F,G\}_\mathrm{f}=\frac{1}{\Gamma}\oint_C\left(\frac{\delta F}{\delta C}(\mathbf{n})\frac{\delta G}{\delta C}(\mathbf{b})-\frac{\delta F}{\delta C}(\mathbf{b})\frac{\delta G}{\delta C}(\mathbf{n})\right)ds,$$

where $\mathbf{n}$ and $\mathbf{b}$ are the unit normal and binormal vector fields on the ring, respectively.

The model generalizes to N non-intersecting rings. The phase space P_f^N is the direct N-product of P_f, minus all points corresponding to intersections of two or more rings, i.e.

$$P_\mathrm{f}^N := \underbrace{(P_\mathrm{f}\times\cdots\times P_\mathrm{f})}_{N\text{ times}}\setminus\triangle. \tag{2.30}$$

The symplectic form and Poisson brackets become summations:

$$\Omega_\mathrm{f}(p)(u,v)=\sum\Gamma_j\oint_{C_j}\langle\mathbf{t}_j(s_j),u(s_j)\times v_j(s_j)\rangle_{\mathbb{R}^3}ds_j,\quad p\in P_\mathrm{f}^N,\quad u,v\in T_pP_\mathrm{f}^N \tag{2.31}$$

$$\{F,G\}_\mathrm{f}(p)=\sum\frac{1}{\Gamma_j}\oint_{C_j}\left\langle\mathbf{t}_j,\left(\frac{\delta F}{\delta C_j}\right)^\sharp\times\left(\frac{\delta G}{\delta C_j}\right)^\sharp\right\rangle_{\mathbb{R}^3}ds_j,$$
$$\frac{\delta F}{\delta C_j},\frac{\delta G}{\delta C_j}\in T_p^*P_\mathrm{f}^N. \tag{2.32}$$

As for point vortices, the Hamiltonian function of the model, $H : P_{\mathrm{f}}^{N} \to \mathbb{R}$, is the kinetic energy due to N rings (the sum of the kinetic energy due to each ring) with the singular terms due to the self-interaction modified as above. This leads to [155][5]

$$H\left(C_{1},\cdots,C_{N}\right)=\sum_{i\neq j}\sum\frac{\Gamma_{j}\Gamma_{i}}{8\pi}\oint_{C_{j}}\oint_{C_{i}}\left\langle \mathbf{t}_{i}(s_{i}),\mathbf{t}_{j}(s_{j})\right\rangle_{\mathbb{R}^{3}}\frac{ds_{i}}{r_{ji}}ds_{j}+\sum H_{\mathrm{SI,reg}}^{j}, \tag{2.33}$$

where $H_{\mathrm{SI,reg}}^{j}$ is $K.E._{\mathrm{SI,reg}}$ for the jth ring and $r_{ji}=\|\mathbf{R}_{ji}\|$ with $\mathbf{R}_{ji}$ denoting the position vector in $\mathbb{R}^3$ of the point (x_j,y_j,z_j) with respect to (x_i,y_i,z_i). Relative to the Marsden–Weinstein symplectic structure (2.31), one then obtains the evolution equations of N vortex rings in $\mathbb{R}^3$ as

$$\frac{\partial C_{j}}{\partial t}=\sum_{i\neq j}\frac{\Gamma_{i}}{8\pi}\oint_{C_{i}}\left(\frac{\mathbf{t}_{i}(s_{i})\times\mathbf{R}_{ji}}{r_{ji}^{3}}\right)_{n_{j}}ds_{i}+\frac{1}{\Gamma_{j}}\left(\frac{\delta H_{\mathrm{SI,reg}}^{j}}{\delta C_{j}}\right)^{\sharp}\times\mathbf{t}_{j},\qquad j=1,\ldots,N. \tag{2.34}$$

In the integral term on the right, the terms within the parentheses represent a vector based on the curve C_j at the point where the curve velocity is being evaluated, and the subscript n_j denotes the component of this vector projected onto the normal plane of C_j at that point (the projection operator is written a few lines after Eq. (4.21) in Chap. 4). If the method of regularization used is the LIA, then one obtains from the last term on the right the binormal self-induced velocity field. However, a word of caution is in order in applying the LIA in this model of mutual interaction of rings. The neglected $O(1)$ terms in (2.25) could be of the same order as the mutual induction terms due to the other rings, and hence, these $O(1)$ terms must be included.[6]

The Hamiltonian form of (2.34) is more easily seen when it is noted that

$$\left(\frac{\delta H}{\delta C_{j}}\right)^{\sharp}(s_{j})=\mathbf{t}_{j}(s_{j})\times\Gamma_{j}\sum_{i\neq j}\frac{\Gamma_{i}}{8\pi}\oint_{C_{i}}\left(\frac{\mathbf{t}_{i}(s_{i})\times\mathbf{R}_{ji}}{r_{ji}^{3}}\right)_{n_{j}}ds_{i}.$$

A derivation of this relation may be found in [155]. Using (2.33) it follows that (2.34) can be written as

$$\frac{\partial C_{j}}{\partial t}=\frac{1}{\Gamma_{j}}\left(\frac{\delta H}{\delta C_{j}}\right)^{\sharp}\times\mathbf{t}_{j},\qquad j=1,\ldots,N. \tag{2.35}$$

This problem may be viewed as a special case of the problem of N rings dynamically interacting in the presence of a rigid boundary, stationary or moving, which will be discussed in more detail in a later chapter.

5 A more detailed derivation of the kinetic energy, with the body included, is presented in a later chapter.

6 I am thankful to Boris Khesin for making this observation.

SE(3) *Symmetry and Momentum Maps*

There is a canonical diagonal SE(3)-action on P_{f}^N, which leaves the Hamiltonian function (2.33) invariant (assuming that $H^j_{\mathrm{SI,reg}}$ also possesses this symmetry; this is true, for example, when the regularization method is the LIA).

Noether's theorem implies that there is a conserved momentum map $\mathbf{J} : P_{\mathrm{f}}^N \to \mathfrak{se}(3)^*$ given by

$$\mathbf{J}(p) = \left(-\frac{1}{2}\sum \Gamma_j \oint_j \langle C_j(s_j), C_j(s_j)\rangle_{\mathbb{R}^3} \mathbf{t}_j(s_j)\, ds_j, \frac{1}{2}\sum \Gamma_j \oint_j (C_j(s_j) \times \mathbf{t}_j)\, ds_j \right). \tag{2.36}$$

Moreover, unlike in the N-point-vortex model, this momentum map is Ad*-equivariant, a complete derivation of this result may be found in [155]. This is an important property of the momentum map that has implications in the symmetry reduction process for Hamiltonian systems; definition and details may be found in [106, 109].

Euler–Lagrange Equations for the N Vortex Ring Model

It is shown in the following pages that Chapman's degenerate Lagrangian can be extended to the N vortex rings model.

Proposition 2.1. *Equations (2.35) are the Euler–Lagrange equations for the following Lagrangian, $\mathscr{L}_{\mathrm{f}} : TP_{\mathrm{f}}^N \to \mathbb{R}$, analogous to Chapman's degenerate Lagrangian for point vortices (2.20),*

$$\mathscr{L}_{\mathrm{f}}\left(C_1, \cdots, C_N, \frac{\partial C_1}{\partial t}, \cdots, \frac{\partial C_N}{\partial t}\right) := -\frac{1}{3}\sum \Gamma_j \oint_{C_j} \left\langle \mathbf{t}_j(s_j), C_j(s_j) \times \frac{\partial C_j}{\partial t}(s_j) \right\rangle_{\mathbb{R}^3} ds_j - H(C_1, \cdots, C_N), \tag{2.37}$$

where H is given by (2.33).

Proof. Define the analog of the quantity V in (2.19) for vortex rings as

$$V_{\mathrm{f}}\left(C_1, \cdots, C_N, \frac{\partial C_1}{\partial t}, \cdots, \frac{\partial C_N}{\partial t}\right) := \sum \Gamma_j \oint_{C_j} \left\langle \mathbf{t}_j(s_j), C_j(s_j) \times \frac{\partial C_j}{\partial t}(s_j) \right\rangle_{\mathbb{R}^3} ds_j$$

and note that the Lagrangian is therefore $\mathscr{L}_{\mathrm{f}} = (-V_{\mathrm{f}}/3) - H$.

To derive the Euler–Lagrange equations, implement the least action principle,

$$\delta \int_{t_1}^{t_2} \mathscr{L}_{\mathrm{f}}\left(C_1, \cdots, C_N, \frac{\partial C_1}{\partial t}, \cdots, \frac{\partial C_N}{\partial t}\right) dt = 0,$$

for variations of the curves, $(\delta C_1(s_i,t),\cdots,\delta C_N(s_N,t)) \in TP_{\mathrm{f}}^N, t \in [t_1,t_2]$, satisfying the fixed end-points condition:

$$\delta C_1(s_1,t_1) = \delta C_1(s_1,t_2) = 0, \cdots, \delta C_N(s_N,t_1) = \delta C_N(s_N,t_2) = 0.$$

As before, each $\delta C_j(s_j,t)$ is identified with a normal vector field based on the jth filament at its instantaneous position at time t.

Consider first the variation in the integral term. Note that there are no variations considered in the differential of the arc-length since the arc-length is not a Lagrangian (or material) parameter.

$$\begin{aligned}
&\delta \int_{t_1}^{t_2} \sum \Gamma_j \oint_{C_j} \left\langle \mathbf{t}_j(s_j,t), C_j(s_j,t) \times \frac{\partial C_j}{\partial t}(s_j,t) \right\rangle_{\mathbb{R}^3} ds_j\, dt \\
&= \sum \Gamma_j \int_{t_1}^{t_2} \oint_{C_j} \left\langle \delta\mathbf{t}_j(s_j,t), C_j(s_j,t) \times \frac{\partial C_j}{\partial t}(s_j,t) \right\rangle_{\mathbb{R}^3} ds_j\, dt \\
&\qquad + \sum \Gamma_j \int_{t_1}^{t_2} \oint_{C_j} \left\langle \mathbf{t}_j(s_j,t), \delta\left(C_j(s_j,t) \times \frac{\partial C_j}{\partial t}(s_j,t) \right) \right\rangle_{\mathbb{R}^3} ds_j\, dt.
\end{aligned} \tag{2.38}$$

To resolve the first set of nested integrals on the right, use the commutability of variations and the use of arc-length parametrization to make the identification

$$\delta \mathbf{t}_j \equiv \frac{\partial(\delta C_j)}{\partial s_j}.$$

And since

$$\oint_{s_j} \frac{\partial}{\partial s_j} \left\langle \delta C_j(s_j,t), C_j(s_j,t) \times \frac{\partial C_j}{\partial t}(s_j,t) \right\rangle_{\mathbb{R}^3} ds_j = 0,$$

it follows that

$$\begin{aligned}
&\int_{t_1}^{t_2} \oint_{C_j} \left\langle \delta\mathbf{t}_j(s_j,t), C_j(s_j,t) \times \frac{\partial C_j}{\partial t}(s_j,t) \right\rangle_{\mathbb{R}^3} ds_j\, dt \\
&= -\int_{t_1}^{t_2} \oint_{C_j} \left\langle \delta C_j(s_j,t), \frac{\partial}{\partial s_j}\left(C_j(s_j,t) \times \frac{\partial C_j}{\partial t}(s_j,t) \right) \right\rangle_{\mathbb{R}^3} ds_j\, dt, \\
&= -\int_{t_1}^{t_2} \oint_{C_j} \left\langle \delta C_j(s_j,t), \mathbf{t}_j(s_j,t) \times \frac{\partial C_j}{\partial t}(s_j,t) \right\rangle_{\mathbb{R}^3} ds_j\, dt \\
&\qquad - \int_{t_1}^{t_2} \oint_{C_j} \left\langle \delta C_j(s_j,t), C_j(s_j,t) \times \frac{\partial \mathbf{t}_j}{\partial t}(s_j,t) \right\rangle_{\mathbb{R}^3} ds_j\, dt.
\end{aligned}$$

Proceeding to the second set of nested integrals on the right of (2.38), use the commutability of variations and derivatives and interchange the sequence of integrals, to write it as

$$\begin{aligned}
&\int_{t_1}^{t_2}\oint_{C_j}\left\langle \mathbf{t}_j(s_j,t),\delta C_j(s_j,t)\times\frac{\partial C_j}{\partial t}(s_j,t)+C_j(s_j,t)\times\frac{\partial(\delta C_j)}{\partial t}(s_j,t)\right\rangle_{\mathbb{R}^3} ds_j\,dt\\
&=\int_{t_1}^{t_2}\oint_{C_j}\left\langle \delta C_j(s_j,t),\frac{\partial C_j}{\partial t}(s_j,t)\times\mathbf{t}_j(s_j,t)\right\rangle_{\mathbb{R}^3} ds_j\,dt\\
&\qquad+\oint_{C_j}\int_{t_1}^{t_2}\left\langle \mathbf{t}_j(s_j,t),C_j(s_j,t)\times\frac{\partial(\delta C_j)}{\partial t}(s_j,t)\right\rangle_{\mathbb{R}^3} dt\,ds_j.
\end{aligned}$$

To resolve the nested integral term in the line above, use the fixed end-points condition to obtain

$$\begin{aligned}
&\int_{t_1}^{t_2}\frac{\partial}{\partial t}\langle \mathbf{t}_j(s_j,t),C_j(s_j,t)\times\delta C_j(s_j,t)\rangle_{\mathbb{R}^3}\,dt\\
&=\langle \mathbf{t}_j(s_j,t_2),C_j(s_j,t_2)\times\delta C_j(s_j,t_2)\rangle_{\mathbb{R}^3}-\langle \mathbf{t}_j(s_j,t_1),C_j(s_j,t_1)\times\delta C_j(s_j,t_1)\rangle_{\mathbb{R}^3}\\
&=0,
\end{aligned}$$

from which it follows that

$$\begin{aligned}
&\int_{t_1}^{t_2}\left\langle \mathbf{t}_j(s_j,t),C_j(s_j,t)\times\frac{\partial(\delta C_j)}{\partial t}(s_j,t)\right\rangle_{\mathbb{R}^3} dt\\
&=-\int_{t_1}^{t_2}\left\langle \frac{\partial \mathbf{t}_j(s_j,t)}{\partial t},C_j(s_j,t)\times\delta C_j(s_j,t)\right\rangle_{\mathbb{R}^3} dt\\
&\qquad-\int_{t_1}^{t_2}\left\langle \mathbf{t}_j(s_j,t),\frac{\partial C_j}{\partial t}(s_j,t)\times\delta C_j(s_j,t)\right\rangle_{\mathbb{R}^3} dt.
\end{aligned}$$

Collecting all these terms, one has

$$\begin{aligned}
&\delta\int_{t_1}^{t_2}\sum\Gamma_j\oint_{C_j}\left\langle \mathbf{t}_j(s_j,t),C_j(s_j,t)\times\frac{\partial C_j}{\partial t}(s_j,t)\right\rangle_{\mathbb{R}^3} ds_j\,dt\\
&\qquad=-\sum 3\Gamma_j\int_{t_1}^{t_2}\oint_{C_j}\left\langle \delta C_j(s_j,t),\mathbf{t}_j(s_j,t)\times\frac{\partial C_j}{\partial t}(s_j,t)\right\rangle_{\mathbb{R}^3} ds_j\,dt.
\end{aligned}\tag{2.39}$$

The least action principle integral finally takes the form

$$\begin{aligned}
&\int_{t_1}^{t_2}\Bigg(\sum\Gamma_j\oint_{C_j}\left\langle \delta C_j(s_j,t),\mathbf{t}_j(s_j,t)\times\frac{\partial C_j}{\partial t}(s_j,t)\right\rangle_{\mathbb{R}^3} ds_j\\
&\qquad-\sum\left\langle \delta C_j(s_j,t),\left(\frac{\delta H}{\delta C_j}\right)^{\sharp}(s_j,t)\right\rangle_{\mathbb{R}^3}\Bigg)\,dt=0.
\end{aligned}$$

Since each δC_j is varied independently, the Euler–Lagrange equations are obtained as

$$\Gamma_j\left(\mathbf{t}_j \times \frac{\partial C_j}{\partial t}\right) = \left(\frac{\delta H}{\delta C_j}\right)^\sharp, \quad j = 1, \cdots, N,$$
$$\Rightarrow \frac{\partial C_j}{\partial t} = \left(\frac{\delta H}{\delta C_j}\right)^\sharp \times \mathbf{t}_j \tag{2.40}$$

keeping in mind that both $\partial C_j/\partial t$ and $(\delta H/\partial C_j)^\sharp$ are both curve-normal vector fields. These equations are identical to (2.35) and this proves the Proposition.

Using (2.29), obtain the canonical momenta from (2.37) as

$$\mathfrak{X} := (\Lambda_1, \cdots, \Lambda_N),$$

where

$$\Lambda_j(s_j) := -\frac{1}{3}\Gamma_j \mathbf{t}_j(s_j) \times C_j(s_j), \quad j = 1, \cdots, N,$$

where each Λ_j, as defined, is a normal vector field based on C_j and $\mathfrak{X}$ represents a family of N such vector fields. As in the N-point-vortex model, this can be formally viewed as an element of the cotangent space by identifying it with a family of the following N normal 1-forms based on C_j:

$$\mathfrak{X}^\flat := \left(\Lambda_1^\flat, \cdots, \Lambda_N^\flat\right), \quad \mathfrak{X}^\flat \in T_p^* P_\mathrm{f}^N,$$

where

$$\Lambda_j^\flat(s_j) := -\left\langle \frac{1}{3}\Gamma_j \mathbf{t}_j(s_j) \times C_j(s_j), dC_j(s_j)\right\rangle_{\mathbb{R}^3} \quad j = 1, \cdots, N.$$

However, again, one finds that the map from $T_p P_\mathrm{f}^N$ to $T_p^* P_\mathrm{f}^N$,

$$\left(\frac{\partial C_1}{\partial t}, \cdots, \frac{\partial C_N}{\partial t}\right)(p) \mapsto \left(\Lambda_1^\flat, \cdots, \Lambda_N^\flat\right)(p),$$

is not invertible being a constant map. It follows that the Lagrangian (2.37) is degenerate. Moreover, $(C_1, \cdots, C_N)$ and $(\Lambda_1^\flat, \cdots, \Lambda_N^\flat)$ are not independent and the equations of motion cannot be cast in canonical Hamiltonian form using the cotangent bundle symplectic structure.

Chapter 3
Dynamically Coupled Rigid Body+Point Vortices in $\mathbb{R}^2$

In this chapter, the extension of the planar Kirchhoff model by the addition of point vortices is discussed. Throughout this chapter the body density will be assumed to be equal to unity. Alternatively, a constant value of $\rho(\neq 1)$ could be used to multiply all mass, momentum, and energy terms. A schematic sketch of the configuration is shown in Fig. 3.1.

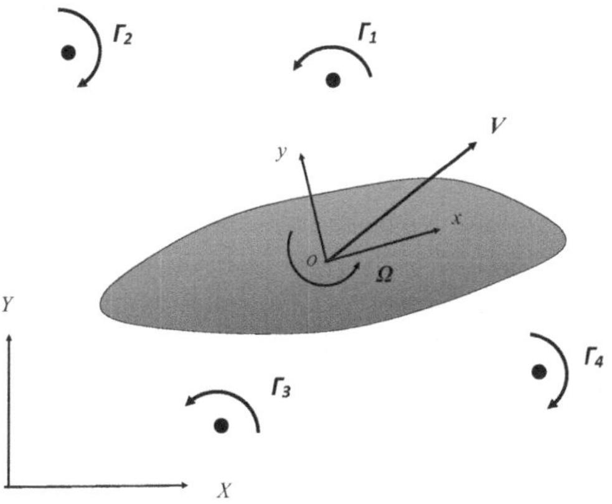

Fig. 3.1 A rigid contour of arbitrary smooth shape, depicting a neutrally buoyant rigid body in the plane, dynamically interacts with four point vortices of arbitrary strengths and signs external to it

Though this chapter mainly follows the approaches used in [150, 156], there are multiple ways in which the governing equations of this system may be derived and formulated, and the reader is also referred to the papers of A. V. Borisov, I. S. Mamaev and S. M. Ramodanov [23–25, 133], in which different approaches and a different set of phase space variables are used. In particular, use is made of complex variable techniques. Reconciliations of these different approaches are discussed in [150] and, from a more technical symmetry reduction perspective, in [173]. To the best of the author's knowledge, the first paper to address this problem is the one by

B. N. Shashikanth, *Dynamically Coupled Rigid Body-Fluid Flow Systems*,
https://doi.org/10.1007/978-3-030-82646-8_3

Jair Koiller [85]. He, however, considered a special limiting case of this problem, one in which the contour is circular and its radius goes to zero. For the case of non-zero circulation around the body, which is not considered in this book, see [25, 174].

Notation

Despite the presence of additional terms that contribute to the system momenta in this and later models, the same symbols, $\mathbf{L}$ and $\mathbf{A}$, that were used in Kirchhoff's model, will be used. This is done to avoid introducing new symbols in each model, and in cases where there is a possibility of confusion the meaning of the variables will be clarified.

Image Flow Fields

In what follows use is made of the classical notion of *image flow fields*, and this is briefly recalled here. When a rigid boundary ∂B is introduced in the domain of an inviscid, incompressible flow with vorticity, the no-penetration boundary condition will in general not be satisfied. The image flow field refers to the velocity flow field, $\mathbf{v}_I$, that must be added such that this boundary condition is satisfied. By definition, this flow field is the solution to the following Neumann problem:

$$\mathbf{v}_I \cdot \mathbf{n} = -\mathbf{v}_o \cdot \mathbf{n} \text{ on } \partial B, \quad \mathbf{v}_I = \nabla \Phi_I, \quad \nabla^2 \Phi_I = 0 \text{ in } D, \tag{3.1}$$

where $\mathbf{v}_o$ is the velocity field of the unperturbed flow. In multiply-connected domains, one allows for the possibility of the image field to have non-zero circulation around boundaries in which case Φ_I is multiple-valued.[1] The important feature of the image field, which follows from the above definition, is that it depends only on the vorticity distribution relative to the location of the rigid boundary. Moreover, the image field itself can in many cases be associated with a distribution of vorticity in the interior or/and on the surface of the body, called the *image vorticity*, and it follows from the previous sentence that this image vorticity does not evolve in time as per the laws of vortex dynamics. This concept has proved to be (computationally) useful and insightful in inviscid flows with vorticity. In particular, for a single rigid body of simple geometry, there are some famous results such as Milne–Thomson's circle theorem [117], Weiss' sphere theorem [184], and the result due to Lewis [97]; the latter two are discussed in Chap. 4.

Sign Convention for the Streamfunction

It is important to note that in this chapter an older convention for the streamfunction is used [92, 117], i.e. velocity components (u, v) are related to the streamfunction ψ by the equations $u = -\partial\psi/\partial y$ and $v = \partial\psi/\partial x$. Equivalently, representing the

[1] And caution must be exercised in evaluating integrals involving Φ_I; the use of branch cuts may be necessary.

streamfunction as a vector field perpendicular to the plane, $\psi\mathbf{k}$, $\mathbf{v} = -\nabla \times \psi\mathbf{k}$. This is also the convention followed in the papers [101, 150, 156].

3.1 N-Point-Vortices and Stationary Rigid Boundaries: C. C. Lin's Problem

Before discussing the dynamically coupled problem with point vortices, it is useful to discuss a precursor problem considered by C. C. Lin—the problem of obtaining the streamfunction for the flow due to N-point-vortices dynamically interacting in a multiply-connected domain $D \subset \mathbb{R}^2$ with *stationary* rigid impenetrable boundaries [101]. The rigid boundaries are a disjoint union of interior boundaries combined with either no exterior boundary or one that is closed or extends to infinity. Obtaining this function also gives the Hamiltonian function for this point vortex system. In Lin's setting, the SE(2) symmetries are lost due to the presence of the fixed boundaries, and in general the three quantities in (2.16) are no longer conserved. But these symmetries are recovered in the dynamically coupled setting.

Using the theory of Green's functions in bounded domains, Lin showed that the motion of the vortices is governed by the equations:

$$\Gamma_j \frac{d\mathbf{r}_j}{dt} = J\frac{\partial W}{\partial \mathbf{r}_j}, \quad j = 1, \cdots, N, \tag{3.2}$$

where

$$W(\mathbf{r}_j) = \sum \Gamma_j \psi_0(\mathbf{r}_j, t) + \sum_{i,j(j>i)} \Gamma_i \Gamma_j G(\mathbf{r}_i; \mathbf{r}_j) + \frac{1}{2}\sum \Gamma_j^2 g(\mathbf{r}_j; \mathbf{r}_j), \tag{3.3}$$

is the Hamiltonian function, labeled the *Kirchhoff–Routh* function by him. In the above, G is a Green's function for the Laplacian operator satisfying a reciprocity property and appropriate boundary and infinity conditions, and of the form

$$G(x, y; x_j, y_j) = g(x, y; x_j, y_j) + (1/4\pi)\log[(x - x_j)^2 + (y - y_j)^2]. \tag{3.4}$$

In fluid dynamic terms, G is the streamfunction of a point vortex of unit strength located at (x_j, y_j) satisfying the no-penetration boundary condition at each point of the stationary rigid boundaries, i.e.

$$G = A_k = \text{constant on } \partial D_k, \quad k = 1, \cdots, M \tag{3.5}$$

(k representing each disjoint component of the interior boundary), in addition to a zero circulation condition on each ∂D_k and appropriate decay conditions at infinity (if unbounded externally); see Lin's papers for details. The logarithmic term is the streamfunction of a unit point vortex in an unbounded flow. The function g is harmonic everywhere in the flow domain and can be identified with the streamfunction of the image flow field added to the logarithmic term to satisfy the boundary conditions. The zero circulation condition for G implies the same condition for g as well. ψ_0 is the streamfunction of any residual, irrotational flow after subtracting the flow due to all the point vortices. W is an invariant of the motion, and Lin asserts that it is equal to the kinetic energy of the flow (modulo the constant density of the fluid). However, as in N-point-vortex models, this identification can be made only after subtracting certain terms that are infinite [15].

In the absence of the residual flow, this assertion may be proved as follows. Consider the case first in which there is a rigid external boundary S enclosing all the vortices and M interior boundaries.

The streamfunction due to a point vortex of strength Γ_j is

$$\psi_j(\mathbf{r},\mathbf{r}_j) = \frac{\Gamma_j}{2\pi} \log \mid \mathbf{r} - \mathbf{r}_j \mid, \tag{3.6}$$

and that due to N-point-vortices is simply the above summed over j till N. Note that the strength of the vortex appears only as a multiplicative factor. From (3.1), it is clear that these same features apply to the image flow field corresponding to each point vortex. Denoting by $\psi^u_{I,j}(\mathbf{r})$ the image streamfunction due to a point vortex of *unit* strength located at $\mathbf{r}_j$, the streamfunction due to N-point-vortices in the multiply-connected domain D is

$$\psi(\mathbf{r},\mathbf{r}_j) = \sum \left(\frac{\Gamma_j}{2\pi} \log \mid \mathbf{r} - \mathbf{r}_j \mid + \Gamma_j \psi^u_{I,j}(\mathbf{r}) \right). \tag{3.7}$$

Applying (2.7) in D, with $\mathbf{A} = \psi \mathbf{k}$, $\omega = \omega \mathbf{k}$, gives

$$\frac{1}{2}\int_D \langle \mathbf{v},\mathbf{v}\rangle_{\mathbb{R}^2}\, \mu = \frac{1}{2}\int_D \psi\omega\, \mu - \frac{1}{2}\int_{\partial D} \psi\mathbf{k}\cdot\mathbf{v}\times\mathbf{n}\, \nu. \tag{3.8}$$

The boundary integral is zero on each of the inner boundaries. On the external boundary S, however, it must be remembered the circulation gets fixed by Stokes theorem applied to D:

$$\oint_S \mathbf{v}\cdot d\mathbf{r} + \sum \Gamma_j = 0. \tag{3.9}$$

And so

$$\frac{1}{2}\oint_S \psi\mathbf{k}\cdot\mathbf{v}\times\mathbf{n}\, \nu = \frac{\psi_{|S}}{2}\int_S \mathbf{k}\cdot\mathbf{v}\times\mathbf{n}\, \nu = \frac{\psi_{|S}}{2}\oint_S \mathbf{v}\cdot d\mathbf{r}.$$

If $\sum\Gamma_j = 0$, then the integral equals zero, and if $\sum\Gamma_j \neq 0$ then recalling that streamfunctions are defined only upto additive constants, one can define a new streamfunction $\tilde{\psi} = \psi - \psi_{|S}$ to make the integral vanish. And so, without loss of generality,

$$\frac{1}{2}\int_D \langle \mathbf{v},\mathbf{v}\rangle_{\mathbb{R}^2}\, \mu = \frac{1}{2}\int_D \psi\omega\, \mu.$$

Substituting ω given by (2.10), subtracting the self-interaction terms, and making the identification $\psi^{u}_{I,j}(\mathbf{r}) \equiv g(\mathbf{r};\mathbf{r}_j)$ prove that the kinetic energy is equal to W given by (3.3) (after setting $\psi_0 = 0$).

In the absence of an external boundary, one has $\psi(\mathbf{r},t) \rightarrow \psi_{|\infty}$, as $|\mathbf{r}| \rightarrow \infty$. Considering an imaginary external boundary C_R, say a circle of radius R, it follows that

$$\lim_{R\to\infty} \frac{1}{2}\oint_{C_R} \psi \mathbf{k}\cdot\mathbf{v}\times\mathbf{n}\, v = \frac{\psi_{|\infty}}{2}\oint_{C_R} \mathbf{v}\cdot d\mathbf{r}.$$

However, from (3.6), it is clear that $|\psi_{|\infty}| = \infty$, and so to make the identification with the kinetic energy this infinite term also needs to be subtracted when $\sum \Gamma_j \neq 0$.

A word of caution is necessary when applying integral identities like (2.7) to a Dirac delta distribution of vorticity. The integrals should, strictly speaking, be computed as improper integrals. This means the domain of computation should exclude small disc-like or tube-like neighborhoods of the vortices and the integrals must be computed in the limit of these neighborhoods shrinking to zero, as in [15]. In this planar case with point vortices, however, one obtains the same result by directly computing the integral on the unpunctured domain and using the properties of Dirac delta distributions.

To summarize, C. C. Lin's system is a Hamiltonian system on the phase space

$$U^{2N} := \mathbb{R}^{2N}\backslash(\triangle \cup B^N), \tag{3.10}$$

where the excluded set $\triangle \cup B^N$ is the set of all collision points of the vortices with each other and the set of N copies of the body domain, the latter implying the exclusion of collision and penetration of each point vortex with the body. The Hamiltonian function $H : U^{2N} \to \mathbb{R}$ is given by $H = -W$, where W is given by (3.3) and the symplectic form is the same as (2.17).

Remark

The system (3.2) and the Kirchhoff–Routh function (3.3) are written in exactly the same way in which C. C. Lin wrote them. Unfortunately, as a result of adopting the older sign convention for streamfunctions mentioned previously, this results in a small discrepancy with the sign convention that is usually adopted for Hamiltonian systems. As per this modern convention, Lin's equations should be written as $d\mathbf{r}_j/dt = -J\partial W/\partial \mathbf{r}_j$, in which case the W function would have a sign opposite to the one used by Lin.

Euler–Lagrange Formulation for C. C. Lin's Problem

The degenerate Lagrangian construction proposed by Chapman for the N-point-vortex model (2.20) works for C. C. Lin's problem too, the only difference being that the Lagrangian is not a conserved quantity; see also [73]. Defining V in exactly the same way as in (2.19) but noting that it is not a conserved quantity in this problem, one defines the Lagrangian as

$$\mathscr{L}_{CCL}(\mathbf{r}_1,\cdots,\mathbf{r}_N,\dot{\mathbf{r}}_1,\cdots,\dot{\mathbf{r}}_j) := -\frac{1}{2}V(\mathbf{r}_1,\cdots,\mathbf{r}_N,\dot{\mathbf{r}}_1,\cdots,\dot{\mathbf{r}}_N) + W(\mathbf{r}_1,\cdots,\mathbf{r}_N). \tag{3.11}$$

The rest of the construction proceeds in a similar way, and the Euler–Lagrange equations give (3.2). It is easily seen that the same canonical momenta as in the N-point-vortex model are obtained leading to the same degeneracies.

3.2 *N*-Point-Vortices Dynamically Coupled with a Single Rigid Contour of Arbitrary Shape

Though the viewpoint adopted in this book is that of extending Kirchhoff's problem, the dynamically coupled problem with point vortices in the plane can also be viewed as an extension of Lin's problem. External forces in Lin's formulation constrain the rigid boundaries to be stationary, in response to the instantaneous fluid pressure distribution on the surface. If these are now removed, the body is free to move in response to the pressure distribution. A principal feature of Lin's problem that facilitates this extension is that, for given boundary shapes, the functions G and g depend for their definition only on the relative *positions* of the vortices from the rigid boundaries. Being solutions to classical Neumann problems in the flow domain, their dependency on time is purely parametric. In principle, then the extension is completed by analyzing the problem in a body-fixed frame. ψ_0 in Lin's formulation is replaced by ψ_B, the streamfunction due to the Kirchhoff flow generated by the motion of the rigid body. The motion of the point vortices is now influenced by this additional flow.

From the perspective of Lin's problem, in which the domain is multiply-connected, the extension can in principle be carried out for multiple moving rigid boundaries; such an attempt is made in [185]. However, the problem with dynamically interacting multiple rigid boundaries, even in the absence of vortices, is sufficiently challenging. For example, the dynamical interaction of even two rigid bodies could be complex, given the observation of non-integrable behavior, *op. cit.*, in Kirchhoff's model of a single rigid body. A brief discussion of the hydrodynamic N-body problem is presented in Chap. 6. The dynamically coupled models in the rest of the book consider only a single rigid boundary.

From the point of view of velocity fields, the fluid velocity field $\mathbf{v}$ of the dynamically coupled problem can be formulated in terms of a Hodge–Helmholtz decomposition:

$$\mathbf{v} = \mathbf{v}_\omega + \mathbf{v}_B, \tag{3.12}$$

where

$$\mathbf{v}_\omega = \mathbf{v}_o + \mathbf{v}_I, \quad \mathbf{v}_\omega \cdot \mathbf{n} = 0 \text{ on } \partial B, \tag{3.13}$$

is the velocity field due to the instantaneous configuration of the vortices relative to the body, obtained as per Lin's formulation and with the image field satisfying the Neumann problem (3.1) (in this case, $\mathbf{v}_o$ is the velocity field due to the N vortices in the unbounded plane). The Neumann problem satisfied by Φ_B, where $\mathbf{v}_B = \nabla \Phi_B$, is the same as (1.3).

The Case of Zero Total Vortex Strength

Define the *total vortex strength* as

$$\Gamma := \sum \Gamma_j.$$

The following equations are derived in [150, 156]. For the case when $\Gamma = 0$, the equations of motion of a planar rigid contour of arbitrary smooth shape dynamically interacting with N-point- vortices referred to a body-fixed frame are

$$\left(\frac{d}{dt} + \Omega \times\right) \mathbf{L} = 0, \tag{3.14}$$

$$\frac{d\mathbf{A}}{dt} + \mathbf{V} \times \mathbf{L} = 0, \tag{3.15}$$

$$\Gamma_j \left(\frac{d\mathbf{l}_j}{dt} + \Omega \times \mathbf{l}_j + \mathbf{V}\right) = J\left(\frac{\partial W}{\partial \mathbf{l}_j}\right), j = 1, \ldots, N, \tag{3.16}$$

where $\mathbf{V}, \Omega, \mathbf{L}$, and $\mathbf{A}$ denote the same physical quantities as in Kirchhoff's model (1.1), with Ω and $\mathbf{A}$ viewed (like the vorticity ω) as vectors perpendicular to the plane. But now, as per the discussion in Sect. 2.1, the vortices also contribute to the quantities $\mathbf{L}$ and $\mathbf{A}$ and relation (1.12) gets modified to

$$\begin{pmatrix} \mathbf{L} \\ \mathbf{A} \end{pmatrix} = (\mathbf{M_b} + \mathbf{M_a}) \cdot \begin{pmatrix} \mathbf{V} \\ \Omega \end{pmatrix} + \begin{pmatrix} \mathbf{p} \\ \pi \end{pmatrix}, \tag{3.17}$$

where

$$\mathbf{p} = \sum \Gamma_j \mathbf{l}_j \times \mathbf{k} + \oint_{\partial B} \mathbf{l} \times (\mathbf{n} \times \mathbf{v}_\omega)\, ds \tag{3.18}$$

and

$$\pi = -\frac{1}{2} \sum \Gamma_j l_j^2 \mathbf{k} - \frac{1}{2} \oint_{\partial B} l^2 (\mathbf{n} \times \mathbf{v}_\omega)\, ds, \tag{3.19}$$

where $\mathbf{v}_\omega$ is given by (3.13). In the above, $\mathbf{l}$ is the position vector of a point (with $l^2 = |\mathbf{l}|^2 = x^2 + y^2$), $\mathbf{l}_j$ is the position vector of the jth point vortex in the body-fixed frame (with $l_j^2 = |\mathbf{l}_j|^2 = x_j^2 + y_j^2$), and

$$W(\mathbf{l}_1, \cdots, \mathbf{l}_N, \mathbf{V}(t), \Omega(t)) = \sum \Gamma_j \psi_B(\mathbf{l}_j, \mathbf{V}(t), \Omega(t)) + W_G(\mathbf{l}_1, \cdots, \mathbf{l}_N) \tag{3.20}$$

is a Kirchhoff–Routh-like function for this problem, with

$$W_G(\mathbf{l}_1, \cdots, \mathbf{l}_N) = \sum_{k,j(j>k)} \Gamma_k \Gamma_j G(\mathbf{l}_k; \mathbf{l}_j) + \frac{1}{2} \sum \Gamma_j^2 g(\mathbf{l}_j; \mathbf{l}_j) \tag{3.21}$$

being the Kirchhoff–Routh function in C. C. Lin's problem—equation (3.3) (with $\psi_0 = 0$)—and ψ_B given by (1.17).

For a given body shape and given strengths of vortices, the contour integrals in (3.18) and (3.19) are functions of $\mathbf{l}_j, j = 1, \cdots, N$. This follows from (3.1) and (3.13). These functions are related to the Kirchhoff potentials, as will be shown in Sect. 3.2.2. The quantities $\mathbf{p}$ and π are, therefore, independent of $(\mathbf{V}, \Omega)$. The relation of these quantities to the vortical momenta is also discussed in that same section.

Proposition 3.1. ([150]) *The system of Eqs. (3.14), (3.15), and (3.16) with Hamiltonian function*

$$\begin{aligned} &H(\mathbf{L}, \mathbf{A}, \mathbf{l}_1, \cdots, \mathbf{l}_N) \\ &:= -W_G(\mathbf{l}_1, \cdots, \mathbf{l}_N) \\ &\qquad + \frac{1}{2} \left\langle (\mathbf{L}, \mathbf{A}) - (\mathbf{p}, \pi), (\mathbf{M_b} + \mathbf{M_a})^{-1} ((\mathbf{L}, \mathbf{A}) - (\mathbf{p}, \pi)) \right\rangle \end{aligned} \tag{3.22}$$

is a Poisson vector field on the space $P = \mathfrak{se}(2)^* \times U^{2N}$, where U^{2N} is given by (3.10), equipped with the following Poisson bracket. For $F, G \in C^\infty(P)$, define, following (1.47),

$$\{F, G\}_P = \{F_1, G_1\}_{\mathfrak{se}(2)^*} + \{F_2, G_2\}_{\mathrm{pv}}. \tag{3.23}$$

Therefore, if $p(t) = (\mu(t), \mathbf{l}_1(t), \cdots, \mathbf{l}_N(t)) \in P$ is an integral curve of the system, where $\mu(t) = (\mathbf{A}(t), \mathbf{L}(t))$, then as per (1.41), (1.36), and (2.18),

$$\frac{dF}{dt} := \left\langle \nabla_p F, \frac{dp}{dt} \right\rangle = \left\langle \frac{\partial F}{\partial \mu}, \mathrm{ad}^*_{\partial H/\partial \mu} \mu \right\rangle_{\mathfrak{se}(2)} + \sum \frac{1}{\Gamma_j} \left(\frac{\partial F}{\partial x_j} \frac{\partial H}{\partial y_j} - \frac{\partial F}{\partial y_j} \frac{\partial H}{\partial x_j} \right), \tag{3.24}$$

where $(x_j, y_j) \equiv \mathbf{l}_j$ now denote the coordinates of the jth vortex in the body-fixed frame.

Remarks

1. The phase space of the vortices in this problem is U^{2N}. Comparing with the phase space described in Sect. 2.2.1, the excluded set is larger since no point vortex can occupy the fixed domain of the body in the body-fixed frame.
2. The point vortex bracket in the proposition is the same as (2.18). The fact that one gets the same bracket in the body-fixed frame has to do with the SE(2) invariance of (2.18).
3. Using (3.20) and certain reciprocity results, see (3.40), that hold for this problem [150], the evolution equations for the point vortices (3.16) can be written in the canonical Hamiltonian form, consistent with (3.24),

$$\Gamma_j \frac{d\mathbf{l}_j}{dt} = -J\frac{\partial H}{\partial \mathbf{l}_j}, j = 1, \ldots, N,$$

   with H given by (3.22).
4. With reference to the discussion in Sect. 1.5.1, it is important to note that the variables $(\mathbf{L}, \mathbf{A})$ and the variables $\mathbf{l}_j$ can be varied independently of each other. This is easily seen from (3.17). Any variation $\delta \mathbf{l}_j, j = 1, \cdots, N$, with $\delta(\mathbf{L}, \mathbf{A}) = 0$, can be achieved by varying $(\mathbf{V}, \Omega)$ so that $\delta(\mathbf{V}, \Omega) = -(\mathbf{M_b} + \mathbf{M_a})^{-1} \cdot \delta(\mathbf{p}, \pi)$. Similarly, any variation $\delta(\mathbf{L}, \mathbf{A})$, with $\delta \mathbf{l}_j = 0, j = 1, \cdots, N$, can be achieved by varying $(\mathbf{V}, \Omega)$ such that $\delta(\mathbf{V}, \Omega) = -(\mathbf{M_b} + \mathbf{M_a})^{-1} \cdot \delta(\mathbf{L}, \mathbf{A})$.
5. As in the planar Kirchhoff problem, simplifications occur when the rigid contour is a circle. The evolution equation for $\mathbf{A}$ essentially decouples from the other equations, and the dimension of the phase space is reduced by one.
6. For an arbitrary (but smooth) contour shape, the image velocity fields and the Kirchhoff unit potentials may not have closed form expressions. In such a case, Eqs. (3.14)–(3.16) have to be appended by Eq. (3.1) for the image velocity field, at each time instant t, and by Eqs. (1.5) and (1.6) for the unit potentials. The latter two sets of equations would however have to be solved (numerically) only once since the contour shape is invariant in the body-fixed frame.

Conserved Quantities

From Remark 4, one has $\partial \mathbf{L}/\partial \mathbf{l}_j = 0$, and it follows that $\mathscr{C}_1$ given by (1.42) is also a Casimir of the bracket (3.23), with $\mathbf{L}$ now given by (3.17). $\mathscr{C}_2$, as in the planar Kirchhoff problem, is conserved trivially.

Recall that in the planar Kirchhoff problem the phase space can be characterized, using the Casimirs, as a cylinder $S^1 \times \mathbb{R}$. In this case, the phase space has an additional U^{2N} component, but due to the presence of the same Casimirs it can be characterized as a U^{2N}-bundle over the same cylinder. The dynamical trajectories lie on the intersection of the level set of the Hamiltonian and this U^{2N}-bundle.

The Case of Non-zero Total Vortex Strength

As per (3.9), $\Gamma \neq 0$ implies circulation at infinity. In this case, the linear momentum equation (3.14) gets modified to

$$\left(\frac{d}{dt} + \Omega\times\right)\mathbf{L} = \Gamma\mathbf{k}\times\mathbf{V}. \tag{3.25}$$

The other evolution equations and the definitions of **L** and **A** are unaltered. From a geometric mechanics viewpoint, this additional term is related to the lack of Ad^*-equivariance of the momentum map. Here Ad^* denotes the coadjoint action of $\mathrm{SE}(2)$ action on $\mathfrak{se}(2)^*$.

For this system, the sum Poisson bracket structure still holds but with an important difference. The Poisson bracket on $\mathfrak{se}(2)^*$ is not the Lie–Poisson bracket but a modified bracket. This modification results from extending the Lie algebra $\mathfrak{se}(2)$, using central extensions and cocyles, and adjusting the coadjoint action to make the momentum map equivariant. This leads to a new Poisson bracket on $\mathfrak{se}(2)^*$, referred to as the cocycle bracket, given by

$$\{F,G\}_{\mathrm{co}} = \{F,G\}_{\mathfrak{se}(2)^*} + \Gamma\left(\frac{\partial F}{\partial L_y}\frac{\partial G}{\partial L_x} - \frac{\partial F}{\partial L_x}\frac{\partial G}{\partial L_y}\right), \quad F,G \in C^\infty(\mathfrak{se}(2)^*). \tag{3.26}$$

The general theory behind these ideas can be found in [2] and [109].

Proposition 3.2. ([150]) *The system of Eqs. (3.25), (3.15), and (3.16) with Hamiltonian function (3.22) is a Poisson vector field on the space $P = \mathfrak{se}(2)^* \times U^{2N}$ equipped with the following Poisson bracket. For $F,G : P \to \mathbb{R}$, define, following (1.47),*

$$\{F,G\}_P = \{F_1,G_1\}_{\mathrm{co}} + \{F_2,G_2\}_{\mathrm{pv}}. \tag{3.27}$$

And, as per (1.41), (3.26), and (2.18), with the same notation as in the previous Proposition,

$$\frac{dF}{dt} := \left\langle \nabla_p F, \frac{dp}{dt}\right\rangle = \left\langle \frac{\partial F}{\partial \mu}, \mathrm{ad}^*_{\partial H/\partial\mu}\mu\right\rangle + \Gamma\left(\frac{\partial F}{\partial L_y}\frac{\partial H}{\partial L_x} - \frac{\partial F}{\partial L_x}\frac{\partial H}{\partial L_y}\right) + \sum\frac{1}{\Gamma_j}\left(\frac{\partial F}{\partial x_j}\frac{\partial H}{\partial y_j} - \frac{\partial F}{\partial y_j}\frac{\partial H}{\partial x_j}\right).$$

Conserved Quantities

Though one still has the conservation of the Hamiltonian H (3.22), $\mathscr{C}_1$ and $\mathscr{C}_2$ of the Lie–Poisson bracket are no longer Casimirs for the cocycle bracket (3.26), as can be easily verified (though $\langle\mathbf{L},\mathbf{A}\rangle$ is still conserved trivially).

3.2.1 The Euler–Lagrange Equations in a Spatially-Fixed Frame

Building upon the Lagrangian descriptions in Sects. 1.4 and 3.1, a Lagrangian description of this problem in the spatially-fixed frame is now proposed. First, a few preliminaries are established.

Define the configuration space for the problem as

$$Q := \mathrm{SE}(2) \times U^{2N}(t), \tag{3.28}$$

where $U^{2N}(t)$ is defined as in (3.10) but is now a time-varying domain due to the motion of the body. Each configuration of the body+point vortices may thus be identified with the pair (g, p), $g \equiv (\bar{\mathbf{r}}_c, \mathbf{R}) \in \mathrm{SE}(2)$, $p \equiv (\bar{\mathbf{r}}_1, \cdots, \bar{\mathbf{r}}_N)$, where $\bar{\mathbf{r}}_j$ and $\bar{\mathbf{r}}_c$ locate the jth point vortex and the body centroid in a spatially-fixed frame.

Define the Lagrangian $\mathscr{L} : TQ \to \mathbb{R}$ as the sum of the Lagrangian in Kirchhoff's problem—for the planar case—and a Lagrangian modified from Eq. (3.11) in C. C. Lin's problem:

$$\begin{aligned}
&\mathscr{L}\left(\mathbf{R}, \bar{\mathbf{r}}_c, \dot{\mathbf{R}}, \bar{\mathbf{V}}, \bar{\mathbf{r}}_1, \cdots, \bar{\mathbf{r}}_N, \dot{\bar{\mathbf{r}}}_1, \cdots, \dot{\bar{\mathbf{r}}}_N\right) \\
&= \frac{1}{2}\langle \mathbf{V}, \mathbf{L}\rangle_{\mathbb{R}^2} + \frac{1}{2}\langle \Omega, \mathbf{A}\rangle_{\mathbb{R}} - \frac{1}{2}V(\bar{\mathbf{r}}_1, \cdots, \bar{\mathbf{r}}_N, \dot{\bar{\mathbf{r}}}_1, \cdots, \dot{\bar{\mathbf{r}}}_N) \\
&\qquad\qquad + \bar{W}(\bar{\mathbf{r}}_1, \cdots, \bar{\mathbf{r}}_N, \bar{\mathbf{V}}, \bar{\Omega}, \bar{\mathbf{r}}_c, \mathbf{R})),
\end{aligned} \tag{3.29}$$

where $\bar{W}$ is the function in (3.20) written in the variables of the spatially-fixed frame, i.e.

$$\bar{W}(\bar{\mathbf{r}}_1, \cdots, \bar{\mathbf{r}}_N, \bar{\mathbf{V}}, \bar{\Omega}, \bar{\mathbf{r}}_c, \mathbf{R}) = \sum \Gamma_j \bar{\psi}_B(\bar{\mathbf{r}}_j, \bar{\mathbf{V}}, \bar{\Omega}, \bar{\mathbf{r}}_c, \mathbf{R}) + \bar{W}_G(\bar{\mathbf{r}}_1, \cdots, \bar{\mathbf{r}}_N, \bar{\mathbf{r}}_c, \mathbf{R}), \tag{3.30}$$

where

$$\bar{\psi}_B(\bar{\mathbf{r}}_j, \bar{\mathbf{V}}, \bar{\Omega}, \bar{\mathbf{r}}_c, \mathbf{R}) := \psi_B(\mathbf{l}_j, \mathbf{V}(t), \Omega(t)) = \psi_B(\mathbf{R}^{-1} \cdot (\bar{\mathbf{r}}_j - \bar{\mathbf{r}}_c), \mathbf{R}^{-1} \cdot \bar{\mathbf{V}}(t), \mathbf{R}^{-1} \cdot \bar{\Omega}(t)), \tag{3.31}$$

$$\bar{W}_G(\bar{\mathbf{r}}_1, \cdots, \bar{\mathbf{r}}_N, \bar{\mathbf{r}}_c, \mathbf{R}) := W_G(\mathbf{l}_1, \cdots, \mathbf{l}_N) = W_G(\mathbf{R}^{-1} \cdot (\bar{\mathbf{r}}_1 - \bar{\mathbf{r}}_c), \cdots, \mathbf{R}^{-1} \cdot (\bar{\mathbf{r}}_N - \bar{\mathbf{r}}_c))$$

and

$$V(\bar{\mathbf{r}}_1, \cdots, \bar{\mathbf{r}}_N, \dot{\bar{\mathbf{r}}}_1, \cdots, \dot{\bar{\mathbf{r}}}_N) := \sum \Gamma_j \left(\bar{x}_j \frac{d\bar{y}_j}{dt} - \bar{y}_j \frac{d\bar{x}_j}{dt} \right).$$

It is important to note that **L** and **A** are those in Kirchhoff's problem, i.e. given by (1.12). In addition to (1.26), one has the following relation between variations:

$$\begin{aligned}\delta \mathbf{l}_j &= \delta\left(\mathbf{R}^{-1}\right)\cdot(\bar{\mathbf{r}}_j-\bar{\mathbf{r}}_c)+\mathbf{R}^{-1}\cdot\delta\bar{\mathbf{r}}_j-\mathbf{R}^{-1}\cdot\delta\bar{\mathbf{r}}_c, \quad j=1,\cdots,N,\\ &= -\left(\mathbf{R}^{-1}\cdot\delta\mathbf{R}\cdot\mathbf{R}^{-1}\right)\cdot(\bar{\mathbf{r}}_j-\bar{\mathbf{r}}_c)+\mathbf{R}^{-1}\cdot\delta\bar{\mathbf{r}}_j-\mathbf{R}^{-1}\cdot\delta\bar{\mathbf{r}}_c, \quad j=1,\cdots,N.\end{aligned} \tag{3.32}$$

There are several degeneracies in this planar problem: the hat map $\hat{}:\mathbb{R}\to\mathfrak{so}(2)$ takes scalars to 2×2 skew-symmetric matrices and the trace pairing (1.23) becomes equivalent to scalar multiplication. The **R**-action on Ω and $\bar{\Omega}$ is the identity making them identical, as are $\dot{\mathbf{R}}$ and $\hat{\Omega}$, and all four can be identified with the same scalar. For purely formal reasons, however, the $\mathbf{R}\cdot$ symbol will still be retained when denoting the action on angular velocities and angular momenta.

Proposition 3.3. *The Euler–Lagrange equations for the Lagrangian (3.29) are equivalent to (3.14)–(3.16) for the case $\Gamma=0$, and to (3.25), (3.15), and (3.16) for the case $\Gamma\neq0$.*

Proof. Implement the least action principle,

$$\delta\int_{t_1}^{t_2}\mathscr{L}\left(\mathbf{R},\bar{\mathbf{r}}_c,\dot{\mathbf{R}},\bar{\mathbf{V}},\bar{\mathbf{r}}_1,\cdots,\bar{\mathbf{r}}_N,\dot{\bar{\mathbf{r}}}_1,\cdots,\dot{\bar{\mathbf{r}}}_N\right)\,dt=0, \tag{3.33}$$

for variations taken over curves $\delta q(t)\equiv(\delta\mathbf{R}(t),\delta\bar{\mathbf{r}}_c(t),\delta\bar{\mathbf{r}}_j(t))\in T_qQ$ satisfying $\delta\mathbf{R}(t_1)=\delta\mathbf{R}(t_2)=0, \delta\bar{\mathbf{r}}_c(t_1)=\delta\bar{\mathbf{r}}_c(t_2)=0$ and $\delta\bar{\mathbf{r}}_j(t_1)=\delta\bar{\mathbf{r}}_j(t_2)=0$ $(j=1,\cdots,N)$.

Importing terms related to the Kirchhoff problem alone from (1.28), with **L** and **A** as in (1.12), and noting that some terms disappear in this planar case obtain

$$\begin{aligned}&\delta\int_{t_1}^{t_2}\mathscr{L}\left(\mathbf{R},\bar{\mathbf{r}}_c,\dot{\mathbf{R}},\bar{\mathbf{V}},\bar{\mathbf{r}}_1,\cdots,\bar{\mathbf{r}}_N,\dot{\bar{\mathbf{r}}}_1,\cdots,\dot{\bar{\mathbf{r}}}_N\right)\,dt\\ &=\int_{t_1}^{t_2}\Bigg(\left\langle\delta\bar{\mathbf{r}}_c,-\frac{d(\mathbf{R}\cdot\mathbf{L})}{dt}-\frac{1}{2}\frac{\partial V}{\partial\bar{\mathbf{r}}_c}+\frac{\partial\bar{W}}{\partial\bar{\mathbf{r}}_c}\right\rangle_{\mathbb{R}^2}\\ &-\frac{1}{2}\operatorname{trace}\left(\delta\mathbf{R}^T\cdot\mathbf{R}\cdot\widehat{\bar{\mathbf{V}}\times\mathbf{L}}\right)-\frac{1}{2}\operatorname{trace}\left(\delta\mathbf{R}^T\cdot\frac{d(\mathbf{R}\cdot\hat{\mathbf{A}})}{dt}-\delta\mathbf{R}^T\cdot\mathbf{R}\cdot\hat{\Omega}\cdot\hat{\mathbf{A}}\right)\\ &\qquad+\left\langle\frac{\partial\bar{W}}{\partial\mathbf{R}},\delta\mathbf{R}\right\rangle_{\mathrm{SO}(2)}+\left\langle\delta\bar{\mathbf{V}},-\frac{1}{2}\frac{\partial V}{\partial\bar{\mathbf{V}}}+\frac{\partial\bar{W}}{\partial\bar{\mathbf{V}}}\right\rangle_{\mathbb{R}^2}+\left\langle\delta\bar{\Omega},\frac{\partial\bar{W}}{\partial\bar{\Omega}}\right\rangle_{\mathbb{R}}\\ &\qquad+\sum\left\langle\delta\bar{\mathbf{r}}_j,-\frac{1}{2}\frac{\partial V}{\partial\bar{\mathbf{r}}_j}+\frac{\partial\bar{W}}{\partial\bar{\mathbf{r}}_j}\right\rangle_{\mathbb{R}^2}+\sum\left\langle\delta\dot{\bar{\mathbf{r}}}_j,-\frac{1}{2}\frac{\partial V}{\partial\dot{\bar{\mathbf{r}}}_j}+\frac{\partial\bar{W}}{\partial\dot{\bar{\mathbf{r}}}_j}\right\rangle_{\mathbb{R}^2}\Bigg)\,dt.\end{aligned}$$

The derivatives of V and $\bar{W}$ with respect to $\bar{\mathbf{r}}_c, \bar{\mathbf{V}}, \bar{\mathbf{r}}_j$, and $\dot{\bar{\mathbf{r}}}_j$ are easily obtained as

$$\frac{\partial V}{\partial \bar{\mathbf{r}}_c} = 0, \quad \frac{\partial V}{\partial \bar{\mathbf{V}}} = 0, \quad \frac{\partial V}{\partial \bar{\mathbf{r}}_j} = \Gamma_j \dot{\bar{\mathbf{r}}}_j \times \mathbf{k}, \quad \frac{\partial V}{\partial \dot{\bar{\mathbf{r}}}_j} = \Gamma_j \mathbf{k} \times \bar{\mathbf{r}}_j, \; j = 1, \cdots, N \tag{3.34}$$

$$\begin{aligned} \frac{\partial \bar{W}}{\partial \bar{\mathbf{r}}_c} &= -\mathbf{R} \cdot \sum \frac{\partial W}{\partial \mathbf{l}_j}, \\ \frac{\partial \bar{W}}{\partial \bar{\mathbf{V}}} &= \mathbf{R} \cdot \sum \Gamma_j \frac{\partial \psi_B}{\partial \mathbf{V}} = \mathbf{R} \cdot \sum \Gamma_j \eta(\mathbf{l}_j), \\ \frac{\partial \bar{W}}{\partial \bar{\mathbf{r}}_j} &= \mathbf{R} \cdot \frac{\partial W}{\partial \mathbf{l}_j}, \quad j = 1, \cdots, N \\ \frac{\partial \bar{W}}{\partial \dot{\bar{\mathbf{r}}}_j} &= 0, \end{aligned} \tag{3.35}$$

where

$$\frac{\partial W}{\partial \mathbf{l}_j} = \Gamma_j \frac{\partial \psi_B}{\partial \mathbf{l}_j} + \frac{\partial W_G}{\partial \mathbf{l}_j}, \quad j = 1, \cdots, N.$$

The derivatives with respect to $\mathbf{R}$ and $\dot{\mathbf{R}}$ are obtained as follows. Since

$$\bar{W}(\bar{\mathbf{r}}_j, \bar{\mathbf{V}}, \bar{\Omega}, \bar{\mathbf{r}}_c, \mathbf{R}) = W(\mathbf{l}_j, \mathbf{V}, \Omega),$$

variations in $\mathbf{R}$ alone, for example, imply that the variations induced in $\bar{W}$ and W are equal with the latter obtained by using the relations (1.26) and (3.32). This gives

$$\begin{aligned} &\left\langle \frac{\partial \bar{W}}{\partial \mathbf{R}}, \delta \mathbf{R} \right\rangle_{SO(3)} \\ &= \sum \left\langle \frac{\partial W}{\partial \mathbf{l}_j}, \delta \mathbf{l}_j \right\rangle_{\mathbb{R}^2} + \left\langle \sum \Gamma_j \frac{\partial \psi_B}{\partial \mathbf{V}}, \delta \mathbf{V} \right\rangle_{\mathbb{R}^2} + \left\langle \sum \Gamma_j \frac{\partial \psi_B}{\partial \Omega}, \delta \Omega \right\rangle_{\mathbb{R}}, \\ &= \sum \left\langle \frac{\partial W}{\partial \mathbf{l}_j}, -\mathbf{R}^{-1} \cdot \delta \mathbf{R} \cdot \mathbf{R}^{-1} \cdot (\bar{\mathbf{r}}_j - \bar{\mathbf{r}}_c) \right\rangle_{\mathbb{R}^2} + \left\langle \sum \Gamma_j \frac{\partial \psi_B}{\partial \mathbf{V}}, -\mathbf{R}^{-1} \cdot \delta \mathbf{R} \cdot \mathbf{V} \right\rangle_{\mathbb{R}^2} \\ &\qquad + \frac{1}{2} \operatorname{trace} \left\langle \widehat{\sum \Gamma_j \frac{\partial \psi_B}{\partial \Omega}}^T \cdot -\mathbf{R}^{-1} \cdot \delta \mathbf{R} \cdot \hat{\Omega} \right\rangle. \end{aligned}$$

Proceeding as in Sect. 1.4 to manipulate the above terms by introducing the element $\hat{\kappa} := \mathbf{R}^{-1} \cdot \delta\mathbf{R} \in \mathfrak{so}(2)$ and using the properties of skew-symmetric matrices and the Euclidean pairings, one obtains

$$\left\langle \frac{\partial \bar{W}}{\partial \mathbf{R}}, \delta\mathbf{R} \right\rangle_{SO(2)}$$
$$= -\frac{1}{2}\operatorname{trace}\left\langle \delta\mathbf{R}^T \cdot \mathbf{R} \cdot \widehat{\sum \mathbf{l}_j \times \frac{\partial W}{\partial \mathbf{l}_j}} \right\rangle - \frac{1}{2}\operatorname{trace}\left\langle \delta\mathbf{R}^T \cdot \mathbf{R} \cdot \widehat{\mathbf{V} \times \sum \Gamma_j \frac{\partial \psi_B}{\partial \mathbf{V}}} \right\rangle$$
$$+ \frac{1}{2}\operatorname{trace}\left\langle \delta\mathbf{R}^T \cdot \mathbf{R} \cdot \left(\hat{\Omega} \cdot \widehat{\sum \Gamma_j \frac{\partial \psi_B}{\partial \Omega}} - \widehat{\Omega \times \sum \Gamma_j \frac{\partial \psi_B}{\partial \Omega}} \right) \right\rangle,$$

the last term in the last pairing again equal to zero due to the planar domain.

Next, for variations in $\bar{\Omega}$ alone,

$$\left\langle \delta\bar{\Omega}, \frac{\partial \bar{W}}{\partial \bar{\Omega}} \right\rangle_{\mathbb{R}^2} = \left\langle \delta\Omega, \frac{\partial W}{\partial \Omega} \right\rangle_{\mathbb{R}^2},$$
$$\Rightarrow \frac{\partial \bar{W}}{\partial \bar{\Omega}} = \mathbf{R} \cdot \frac{\partial W}{\partial \Omega} = \mathbf{R} \cdot \Gamma_j \sum \frac{\partial \psi_B}{\partial \Omega} = \mathbf{R} \cdot \sum \Gamma_j \kappa(\mathbf{l}_j), \tag{3.36}$$

where κ is identified with a vector perpendicular to the plane (refer to Eq. (1.17) and the discussion following it). Using (1.26), the usual assumption of variations commuting with time derivatives and integration by parts, the specific term in the Lagrangian then becomes

$$\int_{t_1}^{t_2} \left\langle \delta\bar{\Omega}, \frac{\partial \bar{W}}{\partial \bar{\Omega}} \right\rangle_{\mathbb{R}^2} dt \equiv \int_{t_1}^{t_2} \frac{1}{2}\operatorname{trace}\left\langle \delta\hat{\Omega}^T \cdot \widehat{\frac{\partial W}{\partial \Omega}} \right\rangle dt,$$
$$= \int_{t_1}^{t_2} \frac{1}{2}\operatorname{trace}\left\langle \left(\mathbf{R}^{-1} \cdot \delta\dot{\mathbf{R}}\right)^T \cdot \widehat{\frac{\partial W}{\partial \Omega}} \right\rangle dt,$$
$$= -\int_{t_1}^{t_2} \frac{1}{2}\operatorname{trace}\left\langle \delta\mathbf{R}^T \cdot \frac{d}{dt}\left(\mathbf{R} \cdot \widehat{\frac{\partial W}{\partial \Omega}}\right) \right\rangle dt,$$
$$= -\int_{t_1}^{t_2} \frac{1}{2}\operatorname{trace}\left\langle \delta\mathbf{R}^T \cdot \frac{d}{dt}\left(\mathbf{R} \cdot \widehat{\sum \Gamma_j \kappa(\mathbf{l}_j)}\right) \right\rangle dt.$$

Performing some more straightforward integration by parts on the terms paired with $\delta\dot{\mathbf{r}}$ and $\delta\bar{\mathbf{V}}$, one finally obtains the following expression for the variation of the action integral:

$$\delta \int_{t_1}^{t_2} \mathscr{L}\left(\mathbf{R}, \bar{\mathbf{r}}_c, \dot{\mathbf{R}}, \bar{\mathbf{V}}, , \bar{\mathbf{r}}_j, \dot{\bar{\mathbf{r}}}_j\right) dt$$
$$= \int_{t_1}^{t_2} \left(\left\langle \delta\bar{\mathbf{r}}_c, -\frac{d(\mathbf{R} \cdot \mathbf{L})}{dt} - \frac{1}{2}\frac{\partial V}{\partial \bar{\mathbf{r}}_c} + \frac{\partial \bar{W}}{\partial \bar{\mathbf{r}}_c} + \frac{d}{dt}\left(\frac{1}{2}\frac{\partial V}{\partial \bar{\mathbf{V}}} - \frac{\partial \bar{W}}{\partial \bar{\mathbf{V}}} \right) \right\rangle_{\mathbb{R}^2} \right.$$

$$-\frac{1}{2}\operatorname{trace}\left(\delta\mathbf{R}^T\cdot\mathbf{R}\cdot\left(\widehat{\mathbf{V}\times\mathbf{L}}+\frac{d\hat{\mathbf{A}}}{dt}+\widehat{\sum\mathbf{l}_j\times\frac{\partial W}{\partial\mathbf{l}_j}}+\widehat{\mathbf{V}\times\sum\Gamma_j\frac{\partial\psi_B}{\partial\mathbf{V}}}\right)\right.$$
$$\left.-\hat{\Omega}\cdot\widehat{\sum\Gamma_j\frac{\partial\psi_B}{\partial\Omega}}+\hat{\Omega}\cdot\widehat{\sum\Gamma_j\kappa(\mathbf{l}_j)}+\frac{d}{dt}\widehat{\sum\Gamma_j\kappa(\mathbf{l}_j)}\right)$$
$$+\sum\left\langle\delta\bar{\mathbf{r}}_j,\left(-\frac{1}{2}\frac{\partial V}{\partial\bar{\mathbf{r}}_j}+\frac{\partial\bar{W}}{\partial\bar{\mathbf{r}}_j}+\frac{d}{dt}\left(\frac{1}{2}\frac{\partial V}{\partial\dot{\bar{\mathbf{r}}}_j}-\frac{\partial\bar{W}}{\partial\dot{\bar{\mathbf{r}}}_j}\right)\right)\right\rangle_{\mathbb{R}^2}\Bigg)\Bigg)\,dt.$$

Implementing the action principle, i.e. by setting the above equal to zero, and using the various relations in (3.34), (3.35), and (3.36), the following $2N+3$ equations are obtained:

$$-\frac{d}{dt}\left(\mathbf{R}\cdot\mathbf{L}+\mathbf{R}\cdot\sum\Gamma_j\eta(\mathbf{l}_j)\right)-\mathbf{R}\cdot\sum\Gamma_j\frac{\partial\psi_B}{\partial\mathbf{l}_j}-\mathbf{R}\cdot\frac{\partial W_G}{\partial\mathbf{l}_j}=0,$$
$$\mathbf{R}\cdot\left[\frac{d\hat{\mathbf{A}}}{dt}+\frac{d}{dt}\widehat{\sum\Gamma_j\kappa(\mathbf{l}_j)}+\mathbf{V}\times\widehat{\left(\mathbf{L}+\sum\Gamma_j\frac{\partial\psi_B}{\partial\mathbf{V}}\right)}+\widehat{\sum\mathbf{l}_j\times\frac{\partial W}{\partial\mathbf{l}_j}}\right]=0,$$
$$\frac{d}{dt}\left(\frac{1}{2}(\Gamma_j\mathbf{k}\times\bar{\mathbf{r}}_j)\right)=\frac{1}{2}\Gamma_j\dot{\bar{\mathbf{r}}}_j\times\mathbf{k}-\mathbf{R}\cdot\frac{\partial W}{\partial\mathbf{l}_j},\quad j=1,\cdots,N.$$

The last term of the angular momentum equation can be re-written using the equations on the last line as

$$\sum\mathbf{l}_j\times\frac{\partial W}{\partial\mathbf{l}_j}=\sum\mathbf{l}_j\times\left(\Gamma_j\left(\frac{d\mathbf{l}_j}{dt}+\Omega\times\mathbf{l}_j+\mathbf{V}\right)\times\mathbf{k}\right),$$
$$=-\frac{1}{2}\frac{d}{dt}\sum\Gamma_jl_j^2+\mathbf{V}\times\left(\sum\Gamma_j\mathbf{l}_j\times\mathbf{k}\right).$$

The Euler–Lagrange equations finally take the form

$$\frac{d}{dt}\left(\mathbf{R}\cdot\left[\mathbf{L}+\sum\Gamma_j\mathbf{l}_j\times\mathbf{k}+\sum\Gamma_j\eta(\mathbf{l}_j)\right]\right)+\Gamma\mathbf{R}\cdot\mathbf{V}\times\mathbf{k}=0,$$
$$\mathbf{R}\cdot\left[\frac{d}{dt}\left(\mathbf{A}-\frac{1}{2}\sum\Gamma_jl_j^2+\sum\Gamma_j\kappa(\mathbf{l}_j)\right)+\mathbf{V}\times\left(\mathbf{L}+\sum\Gamma_j\mathbf{l}_j\times\mathbf{k}+\sum\Gamma_j\eta(\mathbf{l}_j)\right)\right]=0,$$
$$\mathbf{R}\cdot\left(\frac{d\mathbf{l}_j}{dt}+\Omega\times\mathbf{l}_j+\mathbf{V}\right)=\mathbf{R}\cdot\left(\mathbf{k}\times\left(\Gamma_j\frac{\partial\psi_B}{\partial\mathbf{l}_j}+\frac{\partial W_G}{\partial\mathbf{l}_j}\right)\right),\quad j=1,\cdots,N.\tag{3.37}$$

And these are indeed equivalent to the system (3.14), (3.15), and (3.16) (with (3.14) replaced by (3.25) for the case $\Gamma\neq 0$).

This Euler–Lagrange formulation also clearly shows that (3.17) is the correct choice of momentum variables in the Hamiltonian formulation.

3.2.2 The Vortical Momenta and Reciprocity Relations

Referring to Remark 2 in Sect. 1.2, it is useful to examine the relations between the quantities $\mathbf{p}$ and π, defined by Eqs. (3.18) and (3.19), and the formal definitions of the vortical momenta (the momenta due to the free vortices and the image vorticity) in the body-fixed frame, i.e. the integrals $\int_D \mathbf{v}_\omega \, \mu$ and $\int_D \mathbf{l} \times \mathbf{v}_\omega \, \mu$, respectively. These relations are derived by explicit calculations and presented below. In conjunction with reciprocity relations derived in [150], some more relations are also presented. With a view to applications, this is followed by a physical description of the moments of vorticity and moments of circulation terms in (2.5) and (2.6) and how they contribute to the flow linear and angular momenta.

Proposition 3.4. *The quantities* $\mathbf{p}$ *and* π *defined in Eqs. (3.18) and (3.19) are related to the vortical momenta by the following relations:*

$$\mathbf{p} = \int_D \mathbf{v}_\omega \, \mu + \frac{1}{2} \sum \Gamma_j \mathbf{l}_j \times \mathbf{k} - \lim_{R \to \infty} R^2 \oint_0^{2\pi} (\mathbf{v}_I \cdot \mathbf{e}_\theta) \mathbf{e}_\theta \, d\theta,$$

$$\pi = \mathbf{k} \cdot \int_D \mathbf{l} \times \mathbf{v}_\omega \, \mu, \quad \text{only if } \Gamma = 0.$$

If $\Gamma \neq 0$, *the integral* $\int_D \mathbf{l} \times \mathbf{v}_\omega \, \mu$ *is not finite and the second relation is not valid.*

Proof. Consider the subdomain $D_R \subset D$ with an external circular boundary, C_R, of radius R (large) enclosing all the point vortices and the body and centered at the origin of the body-fixed frame. An application of (2.5) in the body-fixed frame (after heeding the word of caution below equation (2.9)), at a given time instant t, gives for the vortical linear momentum in D_R,

$$\int_{D_R} \mathbf{v}_\omega \, \mu = \int_{D_R} \mathbf{l} \times \omega \, \mu + \oint_{\partial B \cup C_R} \mathbf{l} \times (\mathbf{n} \times \mathbf{v}_\omega) \, \nu. \tag{3.38}$$

Now examine the behavior of the contour integral around C_R in the limit $R \to \infty$, $D_R \to D$. For this, the far-field behavior of $\mathbf{v}_\omega$ is needed. Since

$$\mathbf{v}_\omega = J \cdot \nabla \psi(\mathbf{l}, \mathbf{l}_1, \cdots, \mathbf{l}_N)$$

where ∇ is in the coordinates of the body-fixed frame and ψ is the SE(2)-invariant function given by (3.7), write

$$\mathbf{v}_\omega(\mathbf{l}, \mathbf{l}_1, \cdots, \mathbf{l}_N) = \underbrace{J \cdot \left[\sum \frac{\Gamma_j}{2\pi} \frac{\mathbf{l} - \mathbf{l}_j}{|\mathbf{l} - \mathbf{l}_j|^2} \right]}_{\mathbf{v}_o(\mathbf{l}, \mathbf{l}_1, \cdots, \mathbf{l}_N)} + \mathbf{v}_I,$$

where $\mathbf{v}_I = \sum \Gamma_j \mathbf{v}^u_{I,j}(\mathbf{l})$ is the total image velocity field. Setting $\mathbf{l} = -R\mathbf{n}$, evaluate the free vortex term in the first sum for large R,

$$\begin{aligned}
\frac{\Gamma_j}{2\pi}\frac{\mathbf{l}-\mathbf{l}_j}{|\mathbf{l}-\mathbf{l}_j|^2} &= -\frac{\Gamma_j}{2\pi R}\frac{\mathbf{n}+\frac{\mathbf{l}_j}{R}}{|\mathbf{n}+\frac{\mathbf{l}_j}{R}|^2},\\
&= -\frac{\Gamma_j}{2\pi R}\frac{\mathbf{n}+\frac{\mathbf{l}_j}{R}}{\left(1+2\frac{\mathbf{n}\cdot\mathbf{l}_j}{R}+\frac{l_j^2}{R^2}\right)},\\
&= -\frac{\Gamma_j}{2\pi}\left(\frac{\mathbf{n}}{R}+\frac{\mathbf{l}_j}{R^2}\right)\left(1-2\frac{\mathbf{n}\cdot\mathbf{l}_j}{R}+\frac{4(\mathbf{n}\cdot\mathbf{l}_j)^2-l_j^2}{R^2}+O\left(\frac{1}{R^3}\right)\right),\\
&= -\frac{\Gamma_j}{2\pi}\left(\frac{\mathbf{n}}{R}+\frac{\mathbf{l}_j-2\mathbf{n}(\mathbf{n}\cdot\mathbf{l}_j)}{R^2}+O\left(\frac{1}{R^3}\right)\right),\\
\Rightarrow \mathbf{v}_o(\mathbf{l},\mathbf{l}_1,\cdots,\mathbf{l}_N) &= \sum\left[\frac{\Gamma_j}{2\pi}\left(\frac{\mathbf{n}}{R}+\frac{\mathbf{l}_j-2\mathbf{n}(\mathbf{n}\cdot\mathbf{l}_j)}{R^2}+O\left(\frac{1}{R^3}\right)\right)\right]\times\mathbf{k}.
\end{aligned}$$

Denote by $\mathbf{e}_\theta$ the unit tangent field on C_R, positive in the counter-clockwise direction implying that $\mathbf{e}_\theta = \mathbf{n}\times\mathbf{k}$. Writing $\nu = Rd\theta$, and using elementary vector identities, it follows that

$$\oint_{C_R} \mathbf{l}\times(\mathbf{n}\times\mathbf{v}_o)\ \nu = -\frac{1}{2}\sum\Gamma_j\mathbf{l}_j\times\mathbf{k}+O\left(\frac{1}{R}\right),$$

where the relation $\int_0^{2\pi}(\mathbf{l}_j\cdot\mathbf{n})(\mathbf{n}\times\mathbf{k})\,d\theta = \pi(\mathbf{l}_j\times\mathbf{k})$ has been used.

To evaluate the far-field behavior of the corresponding integral for the image field, recall from (3.1) that the image field is a potential field everywhere in D. Moreover, since the circulation around the body is assumed to be zero, this field has zero circulation around any closed contour enclosing the body. Such a potential field in the plane has a far-field decay rate given by

$$\mathbf{v}_I = O\left(\frac{1}{R^2}\right), \quad R\to\infty,$$

and so $\oint_{C_R}\mathbf{l}\times(\mathbf{n}\times\mathbf{v}^u_I)\ \nu = O(1), R\to\infty$. Returning to (4.45), it follows that as $R\to\infty$,

$$\begin{aligned}
\int_D \mathbf{v}_\omega\,\mu = \sum\Gamma_j\mathbf{l}_j\times\mathbf{k}+\oint_{\partial B}\mathbf{l}\times(\mathbf{n}\times\mathbf{v}_\omega)\ \nu\\
-\frac{1}{2}\sum\Gamma_j\mathbf{l}_j\times\mathbf{k}+\lim_{R\to\infty}R^2\oint_0^{2\pi}(\mathbf{v}_I\cdot\mathbf{e}_\theta)\mathbf{e}_\theta\,d\theta.
\end{aligned}$$

Note that the integral involving the image fields should be viewed as a body boundary integral since it can be recast in terms of the first moment of image vorticity. Interestingly, therefore, and with reference to Remark 2 in Sect. 1.2, $O(1)$ contributions due to the unbounded nature of the domain can make the variable $\mathbf{p}$ different from the linear vortical momentum.

The computation for the vortical angular momentum is easier. Applying (2.6) to $\mathbf{v}_\omega$ in D_R,

$$\int_{D_R} \mathbf{l}\times\mathbf{v}_\omega\,\mu = -\frac{1}{2}\int_{D_R} l^2\omega\,\mu - \frac{1}{2}\oint_{\partial B\cup C_R} l^2\,(\mathbf{n}\times\mathbf{v}_\omega)\,\nu. \tag{3.39}$$

Consider the behavior of the far-field integral

$$\begin{aligned}\oint_{C_R} l^2\,(\mathbf{n}\times\mathbf{v}_\omega)\,\nu \equiv R^2\oint_{C_R}\mathbf{n}\times\mathbf{v}_\omega\,\nu &= -R^2\int_{D_R}\omega\,\mu - \oint_{\partial B}\mathbf{n}\times\mathbf{v}_\omega\,\nu,\\ &= -R^2\sum\Gamma_j,\end{aligned}$$

since $\mathbf{v}_o$ is irrotational in the domain occupied by the body and with the assumption that $\mathbf{v}_I$ has zero circulation around the body.

Returning to (3.39), it follows that as $R\to\infty$,

$$\begin{aligned}\int_D \mathbf{l}\times\mathbf{v}_\omega\,\mu &= \infty, \quad \text{if}\,\Gamma\neq 0,\\ &= \pi, \quad \text{if}\,\Gamma = 0.\end{aligned}$$

This proves the Proposition.

Another set of interesting relations exists in this problem that holds for any distribution of vortex strengths and any smooth contour shape. These were named as reciprocity relations in [150] (see also [186, 187]):

$$\left(\sum\Gamma_j\eta(\mathbf{l}_j),\sum\Gamma_j\kappa(\mathbf{l}_j)\right) = \left(\oint_{\partial B}\mathbf{l}\times(\mathbf{n}\times\mathbf{v}_\omega)\,\nu, -\frac{\mathbf{k}}{2}\cdot\oint_{\partial B} l^2\,(\mathbf{n}\times\mathbf{v}_\omega)\,\nu\right), \tag{3.40}$$

where η and κ are as defined in (1.17), $\eta\equiv(\eta_1,\eta_2)$.

Substituting these relations in (3.18) and (3.19) and using the relations of Proposition 3.4, another set of interesting relations is obtained

$$\int_D \mathbf{v}_\omega\,\mu = \frac{1}{2}\sum\Gamma_j\mathbf{l}_j\times\mathbf{k} + \sum\Gamma_j\eta(\mathbf{l}_j) + \lim_{R\to\infty} R^2\oint_0^{2\pi}(\mathbf{v}_I\cdot\mathbf{e}_\theta)\,\mathbf{e}_\theta\,d\theta, \tag{3.41}$$

$$\mathbf{k}\cdot\int_D \mathbf{l}\times\mathbf{v}_\omega\,\mu = -\frac{1}{2}\sum\Gamma_j l_j^2 + \sum\Gamma_j\kappa(\mathbf{l}_j), \quad \text{only if}\,\Gamma = 0. \tag{3.42}$$

These relations show explicitly how, for a given body shape, the linear and angular momenta of the flow generated by the point vortices are a function of only their spatial distribution relative to the body and their strengths.

It is useful to get a physical sense of how the spatial distribution of vorticity contributes directly to the momenta of the flow by considering the simple case of a single point vortex of positive strength Γ occupying a position on the y-axis, and with no body in the flow, as shown in Fig. 3.2.[2] Reverting to Eq. (2.5), it is easily seen that the first moment of vorticity term makes a positive contribution to the flow linear momentum in the x-direction and a zero contribution to the flow linear momentum in the y-direction. Some more straightforward calculation shows that the first moment of circulation terms contributes in the same manner, so that the total linear momentum is all along the positive x-axis. To see this directly, consider the linear momentum integral in a domain enclosed by the circle C centered at the origin of the xy-frame, shown by the black dashed curve. Recalling the classical Γ/r velocity field of the vortex, it is easily seen that owing to the circular symmetry of this field the linear momentum is zero in the domain enclosed by the blue dashed circle centered at the location of the vortex. The straight red dashed line denotes the locus of points with the same y-coordinate as the vortex. In the regions enclosed by the black, blue, and red curves, and denoted by N, the velocity vector at each point has an x-component in the negative direction. This whole region therefore makes a contribution in the negative x-direction to the linear momentum. In the region enclosed by the black, blue, and red curves, and denoted by P, the velocity vector at each point has an x-component in the positive direction. The larger size of P thus results in an overall contribution in the positive x-direction to the linear momentum. The argument can be extended to a circle C of any size to show that one gets positive x-component linear momentum in the whole domain. A similar diagram and argument, by drawing the dashed red line along the y-axis, can be used to show why the linear momentum in the y-direction is zero. Consider next the angular momentum

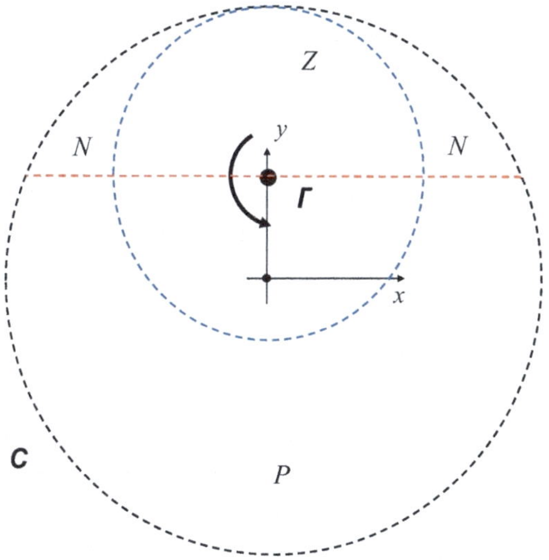

Fig. 3.2 A schematic diagram to explain the moments of vorticity and circulation due to a single point vortex

[2] Needless to add, these intuitive arguments should not be considered a replacement for the actual computations.

of the flow. Reverting to Eq. (2.6), it is easily seen that the second moment of vorticity term makes a negative contribution. However, the second moment of circulation makes a bigger positive contribution so that the net angular momentum of the flow is positive, i.e. the angular momentum is counter-clockwise about the origin. This can also be inferred from Fig. 3.2, where in most of the domains the velocity arrows are directed such as to cause a counter-clockwise moment about the origin.

To highlight the role of the first moment of vorticity term in (3.41) in the transfer of linear momentum from the fluid to the body, consider the problem of the shedding of vortices by swimming fish and other marine creatures to gain forward momentum. These vortices are shed from the various types of fins located along a fish body—the pectoral, dorsal, pelvic, and caudal fins. The latter is the tail fin, and the other smaller fins are located at various locations fore of the caudal. These vortices are created by viscosity, but the conservation of linear momentum for the system "fish+surrounding water" holds for the shedding process, since this is an internal mechanism. Zero change in the system linear momentum implies that the change in the fish body's momentum, post-shedding, is opposite in direction to the change in the vortical momentum of the water.

In Fig. 3.3, the first moment of vorticity vector is shown at various locations of the point vortex relative to the body's center of mass, assumed to be at the origin of the body-fixed frame. For the body to gain linear momentum in the negative x-direction, the best locations for the shed vortex of positive strength are near T or, more generally, closer to the positive y-axis (and if the strength is negative then closer to the negative y-axis and point B). This explains how the smaller body fins in a fish contribute to forward momentum. Moving to the aft portion of the body, a vortex shed close to point A corresponds to a moment of vorticity vector mainly in the y-direction. However, the x-component of this can be made larger by shedding sufficiently downstream and increasing the moment arm thereby increasing the magnitude of the moment of vorticity vector. This explains how the caudal fin contributes to forward motion.

Figure 3.3 also illustrates a general feature about the flow linear momentum generated by a group of vortices in the vicinity of a body. If the configuration defined by these vortices is scaled up by moving each of vortex proportionately in a radially outward manner, then the corresponding first moment of vorticity vectors scales up accordingly, and thereby, from (3.41), there is a corresponding increase in the magnitude of the total linear momentum vector due to these vortices. Therefore as a thumb rule, one can state that as a configuration of vortices approaches a body in a manner in which their moment arms do not significantly change direction, then the flow linear momentum generated by these vortices decreases in magnitude.

Clearly, there could be other mechanisms that contribute to the linear momentum transfer. In the above argument, the contribution of the body shape through the added mass terms and the terms $\Gamma_j \eta(\mathbf{l}_j)$ in Eq. (3.41) has not been considered. And since the framework of this model neglects deformation of the body, viscosity, and three-dimensionality, all these real-world effects must be taken into account to form a complete picture of the momentum transfer mechanism in the swimming of aquatic creatures. Some of these relations will be examined again in three-dimensional and viscous settings with the no-slip boundary condition enforced.

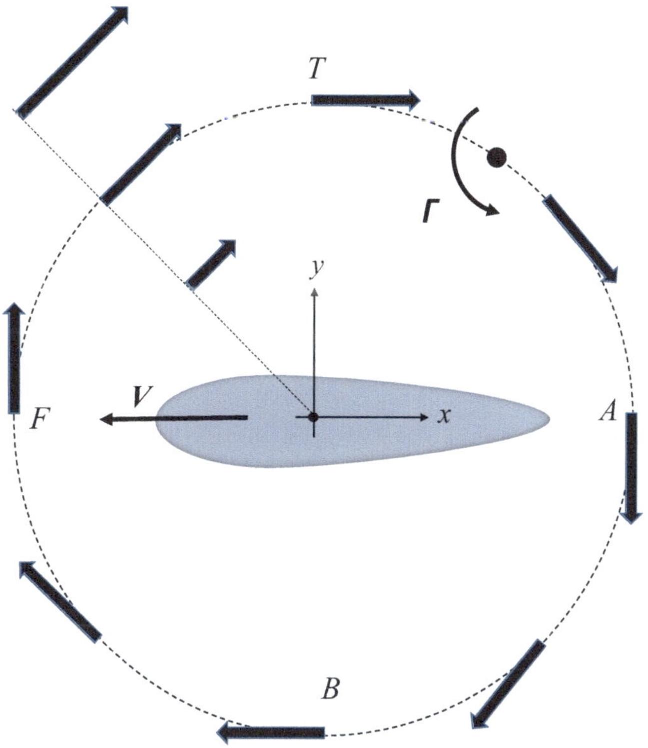

Fig. 3.3 The first moment of vorticity vector, depicted by the thick black arrows, corresponding to different locations of a point vortex of positive strength Γ. For locations on a circle of constant radius centered at the origin of the body-fixed frame, the vector has constant magnitude but varying direction. On a radial line from the origin, the vector has constant direction but increases linearly in magnitude. The letters T, B, F, and A stand for "top," "bottom," "fore," and "aft," respectively

3.3 *N*-Point-Vortices Dynamically Coupled with a Single Rigid Circular Contour

As in the planar Kirchhoff problem, the case of a circular rigid contour offers some simplifications. A schematic sketch is shown in Fig. 3.4. Once again set $\Omega = 0$, without loss of generality. As a consequence of (1.44), $\mathbf{A}$ is no longer a dynamical variable of the system and, H, as given by (3.22), is a function only of $\mathbf{L}$ and $(\mathbf{l}_1, \cdots, \mathbf{l}_N)$. The $\mathbf{A}$ evolution equation in (3.44) is dropped. The phase space dimension for this case therefore effectively reduces by 1 and the phase space can be characterized as $\mathbb{R}^2 \times U^{2N}$. For the case $\Gamma = 0$, Eq. (1.45) still holds and provides a stronger conservation law than $\mathscr{C}_1$ alone. The dynamical trajectories of the system for this case lie on the intersection with $\mathbb{R}^2 \times U^{2N}$ of the level sets of the Hamiltonian and $\mathbf{L}$. For the case $\Gamma \neq 0$, the conservation of $\mathbf{L}$ and $\mathscr{C}_1$ no longer holds, but there exists an S^1-symmetry in the system and a conserved momentum map, as shown later.

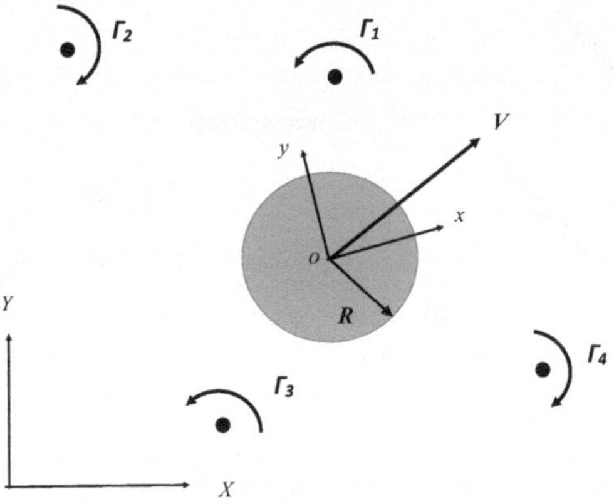

Fig. 3.4 A rigid circle, depicting a neutrally buoyant circular cylinder in the plane, dynamically interacts with four point vortices of arbitrary strengths and signs external to it

Yet another advantage offered by this simple geometry, especially from a computational viewpoint, is that the image velocity field (3.1) has a simple and explicit representation resulting from Milne–Thomson's famous circle theorem [116, 117]. Recall that this theorem states that the image field due to a point vortex of strength Γ_j, external to a rigid circular contour and located at (x_j, y_j) from the center of the circle, is the velocity field obtained by placing point vortices inside the circle at specified locations. There are two choices for the placement of these image vortices and their strengths: (a) a single point vortex of strength $-\Gamma_j$ at the reciprocal point, i.e. the point with coordinates $(x_j, y_j)R^2/(x_j^2 + y_j^2)$ or (b) a pair of point vortices, one at the reciprocal point, as in (a), and the other at the center of the circle and of strength Γ_j. The difference is that (a) has non-zero circulation associated with it and (b) has zero circulation associated with it.

Using the image vortex representation (b), the function g in (3.4) is given by

$$g(x,y;x_j,y_j) = \frac{1}{4\pi}\log\left(x^2+y^2\right) - \frac{1}{4\pi}\log\left((x-R^2x_j/l_j^2)^2 + (y-R^2y_j/l_j^2)^2\right).$$

The Kirchhoff flow of the circle has streamfunction and velocity potential function given, respectively, by

$$\psi_B(\mathbf{l},\mathbf{V}) = R^2\left\langle \left(\frac{-y}{l^2}, \frac{x}{l^2}\right), \mathbf{V}\right\rangle, \quad \Phi_B(\mathbf{l},\mathbf{V}) = -R^2\left\langle \left(\frac{x}{l^2}, \frac{y}{l^2}\right), \mathbf{V}\right\rangle. \tag{3.43}$$

One then obtains the system of equations as [156]:

$$\begin{aligned} \frac{d\mathbf{L}}{dt} &= \Gamma \mathbf{k}\times\mathbf{V}, \\ \Gamma_j \frac{d\mathbf{l}_j}{dt} &= -J\frac{\partial H}{\partial \mathbf{l}_j}, \qquad j = 1, \ldots\ldots, N. \end{aligned} \tag{3.44}$$

In the above,

$$\mathbf{L} = c\mathbf{V} + \sum \Gamma_j \mathbf{l}_j \times \mathbf{k} + R^2 \sum \mathbf{k} \times \Gamma_j \left(\frac{x_j}{l_j^2}, \frac{y_j}{l_j^2} \right), \tag{3.45}$$

where $c = 2\pi R^2$ is the term denoting the mass plus added mass of the 2-D circular cylinder of unit density, and

$$\begin{aligned} H(\mathbf{L},\mathbf{A},\mathbf{l}_1,\cdots,\mathbf{l}_N) &= -W(\mathbf{L},\mathbf{l}_1,\cdots,\mathbf{l}_N) + \frac{1}{2c}\langle \mathbf{L},\mathbf{L}\rangle_{\mathbb{R}^2} \\ &\quad -\frac{1}{c}\left(\sum \Gamma_j(\mathbf{L}\times\mathbf{l}_j)\cdot\mathbf{k} - \frac{1}{2}\sum \Gamma_j^2 l_j^2 - \sum_{i,j(j>i)} \Gamma_i\Gamma_j\langle \mathbf{l}_i,\mathbf{l}_j\rangle \right. \\ &\quad \left. + \frac{R^4}{2}\left\langle \sum \Gamma_j \frac{\mathbf{l}_j}{l_j^2}, \sum \Gamma_j \frac{\mathbf{l}_j}{l_j^2} \right\rangle_{\mathbb{R}^2} \right). \end{aligned} \tag{3.46}$$

$W(\mathbf{L},\mathbf{l}_j)$ is the Kirchhoff–Routh function (3.20) for this system. Equivalently, one can write H in the form of (3.22). The Poisson bracket structure defined by (3.26) and (3.27) on $\mathfrak{se}(2)^* \times U^{2N}$ still formally applies, but one restricts to functions that depend only on $\mathbf{L}$ and $(\mathbf{l}_1,\cdots,\mathbf{l}_N)$. This actually makes the Lie–Poisson bracket trivial, as easily seen from (1.38). Due to these simplifications, alternative Poisson bracket formulations in which there is no Lie–Poisson bracket are also possible. For the case $\Gamma = 0$, by restricting to a level set of $\mathbf{L}$ in the phase space and no longer considering $\mathbf{L}$ as a dynamical variable, one can use just the point vortex bracket on $P = U^{2N}$. For the case, $\Gamma \neq 0$, an alternative Poisson formulation is the sum of the point vortex and the cocycle brackets, i.e. for $P \equiv \mathbb{R}^2 \times U^{2N}$ and $F, G : P \to \mathbb{R}$,

$$\{F,G\}_P = \Gamma\left(\frac{\partial F}{\partial L_y}\frac{\partial G}{\partial L_x} - \frac{\partial F}{\partial L_x}\frac{\partial G}{\partial L_y} \right) + \sum \frac{1}{\Gamma_j}\left(\frac{\partial F}{\partial x_j}\frac{\partial G}{\partial y_j} - \frac{\partial F}{\partial y_j}\frac{\partial G}{\partial x_j} \right). \tag{3.47}$$

This bracket was used in [152] where the dynamics and control for the case of a single vortex were studied.

The S^1-Symmetry and Its Momentum Map

Recall that when $\Gamma \neq 0$, neither $\mathbf{L}$ nor $\mathscr{C}_1$ is a conserved quantity, regardless of the shape of the body. However, for circular shapes there exists a conserved quantity due to the S^1-symmetry. Consider the following diagonal S^1 action on the phase space $P \equiv \mathfrak{se}(2)^* \times U^{2N}$ equipped with the bracket (3.27) (equivalently one can consider the S^1 action on $P \equiv \mathbb{R}^2 \times U^{2N}$ equipped with the bracket (3.47), since $\mathbf{A}$ no longer plays any role in the dynamics). Identifying S^1 with SO(2), define $A : \mathrm{SO}(2) \times P \to P$ by

$$A \cdot (\mathbf{A},\mathbf{L},\mathbf{l}_1,\cdots,\mathbf{l}_N) := (\mathbf{A},\mathbf{R}\cdot\mathbf{L},\mathbf{R}\cdot\mathbf{l}_1,\cdots,\mathbf{R}\cdot\mathbf{l}_N), \quad \mathbf{R} \in \mathrm{SO}(2). \tag{3.48}$$

In the expression for H, (3.46), it is immediately obvious that, except for the first term, each term on the right is invariant under this action. Referring to (3.20), this remaining term can be written explicitly. First, note that from (3.45), the induced action on $\mathbf{V}$ is exactly the same as the above. Next, the individual terms in the various sums in $W(\mathbf{L},\mathbf{l}_j)$ are

$$\Gamma_j \psi_B(\mathbf{l}_j,\mathbf{V}) = R^2 \Gamma_j \frac{\mathbf{l}_j \times \mathbf{V}}{l_j^2},$$

$$\frac{1}{2}\Gamma_j^2 g(\mathbf{l}_j;\mathbf{l}_j) = -\frac{1}{4\pi}\Gamma_j^2 \log\left(1 - \frac{R^2}{l_j^2}\right), \tag{3.49}$$

$$\Gamma_k\Gamma_j G(\mathbf{l}_k;\mathbf{l}_j) = \frac{1}{4\pi}\Gamma_k\Gamma_j \log\left(\frac{l_k^2}{\left\langle \mathbf{l}_k - \frac{R^2}{l_j^2}\mathbf{l}_j, \mathbf{l}_k - \frac{R^2}{l_j^2}\mathbf{l}_j \right\rangle_{\mathbb{R}^2}}\right) + \frac{1}{4\pi}\Gamma_k\Gamma_j \log\left(\langle \mathbf{l}_k - \mathbf{l}_j, \mathbf{l}_k - \mathbf{l}_j \rangle\right). \tag{3.50}$$

Clearly, each term is invariant under the A-action, and the summations do not alter this fact. Therefore, $-W(\mathbf{L},\mathbf{l}_j)$, and hence H given by (3.46), is invariant under the A-action. It follows from Noether's Theorem that the associated momentum map, $\mathbf{J}: P \to \mathbb{R}$, is conserved by the dynamics. Here, $\mathbb{R}$ is identified with $\mathfrak{so}(2)^*$, the dual of the Lie algebra of SO(2).

To compute the momentum map, first compute the infinitesimal generator ξ_P that is a vector field on P corresponding to the infinitesimal action of a Lie algebra element $\xi \in \mathfrak{so}(2) \equiv \mathbb{R}$ (for the general definition, properties, and computation of the infinitesimal generator, see, for example, [109]). At each $p \equiv (\mathbf{A},\mathbf{L},\mathbf{l}_1,\cdots,\mathbf{l}_N) \in P$, it is defined as

$$\xi_P(p) := \frac{d}{dt}\left(A(t)\cdot p\right)_{|t=0} \equiv \frac{d}{dt}\left(\mathbf{A}, \mathbf{R}(t)\cdot\mathbf{L}, \mathbf{R}(t)\cdot\mathbf{l}_1,\cdots,\mathbf{R}(t)\cdot\mathbf{l}_N\right)_{|t=0},$$

where $\mathbf{R}(t)$ is a parametrization of SO(2) that satisfies

$$\mathbf{R}(0) = \mathrm{Id}, \quad \frac{d}{dt}\left(\mathbf{R}(t)\right)_{|t=0} = \xi.$$

Representing $\mathbf{R}$ by the 2×2 matrix

$$\begin{pmatrix} \cos\theta & -\sin\theta \\ \sin\theta & \cos\theta \end{pmatrix},$$

or, equivalently, by just θ so that $d\theta(0)/dt \equiv \xi \in \mathbb{R}$, it follows that

$$\xi_P(p) := (0, J\cdot\mathbf{L}, J\cdot\mathbf{l}_1,\cdots,J\cdot\mathbf{l}_N)\,\xi,$$

where J is given by (2.14).

Next, obtain the function $H_{\mathbf{J}}(\xi) : P \to \mathbb{R}$ for which ξ_P is the Hamiltonian vector field relative to the Poisson bracket structure (3.27). For any $H : P \to \mathbb{R}$, such a Hamiltonian vector field has the form

$$X_H = \left(\mathbf{L} \times \frac{\partial H}{\partial \mathbf{L}}, -\Gamma \frac{\partial H}{\partial L_y}, \Gamma \frac{\partial H}{\partial L_x}, \frac{1}{\Gamma_1} \frac{\partial H}{\partial y_1}, -\frac{1}{\Gamma_1} \frac{\partial H}{\partial x_1}, \dots, \frac{1}{\Gamma_N} \frac{\partial H}{\partial y_N}, -\frac{1}{\Gamma_N} \frac{\partial H}{\partial x_N} \right). \tag{3.51}$$

Comparing the previous two equations, it is easy to see that

$$H_{\mathbf{J}}(\xi) = \frac{1}{2} \left(\frac{\langle \mathbf{L}, \mathbf{L} \rangle}{\Gamma} - \sum \Gamma_j l_j^2 \right) \xi.$$

Recalling that the momentum map satisfies

$$\langle \mathbf{J}(p), \xi \rangle = H_{\mathbf{J}}(\xi)(p),$$

the conserved momentum map $\mathbf{J} : \mathfrak{se}(2)^* \times U^{2N} \to \mathbb{R}$ is obtained as

$$\mathbf{J}(p) = \frac{1}{2} \left(\frac{\langle \mathbf{L}, \mathbf{L} \rangle}{\Gamma} - \sum \Gamma_j l_j^2 \right), \quad \Gamma \neq 0.$$

For the case $\Gamma = 0$, the above is not defined. Since the cocycle bracket disappears for this case, the entries with Γ in X_H are zero, thus making it impossible for ξ_P to be a Hamiltonian vector field. One could restrict the group action to a level set of $\mathbf{L}$ that, as mentioned previously, can be identified with U^{2N} equipped with the point vortex bracket. The entries $J \cdot \mathbf{L}$ in ξ_P are now zero and one can formally obtain a momentum map $\mathbf{J} : U^{2N} \to \mathbb{R}$ as $\mathbf{J}(p) = -\sum \Gamma_j l_j^2 / 2$. However, this action of S^1 does not leave H, (3.46), invariant. This momentum map is therefore *not* a conserved quantity.

The Case $N = 1$

For the case when there is only a single vortex, the four-dimensional phase space $P = U^2 \times \mathbb{R}^2$ and $\Gamma = \Gamma_1$. This system is integrable; indeed, one can perform the Marsden–Weinstein symplectic reduction explicitly and obtain a Hamiltonian system on a two-dimensional symplectic manifold [152]; see also [23]. Two types of vortex orbits relative to the moving cylinder are also considered in [152]—bound orbits, in which the vortex always stays within a finite neighborhood of the cylinder, and scattering orbits in which the vortex approaches the cylinder from infinity and eventually escapes to infinity.

The Case $N = 2$ and the Moving Föppl Equilibrium

For the case when there is a point vortex pair, the phase space, as explained previously, is $P = U^4 \times \mathbb{R}^2$ when $\Gamma := \Gamma_1 + \Gamma_2 \neq 0$ and is $P = U^4$ when $\Gamma = 0$. For the latter, the additional presence of the S^1-momentum map should, in principle, allow one to perform symplectic reduction and obtain a two-dimensional (integrable) Hamiltonian system. For discussions on integrability and nonintegrability in this case, see [24, 139].

The case $\Gamma = 0$ also provides an interesting relative equilibrium (or moving equilibrium) configuration. This is the analog of the famous Föppl equilibrium in which a steady streaming flow over a stationary circular contour and a symmetrically located counter-rotating point vortex pair are in equilibrium [46, 147]. Hill further extended these equilibrium solutions to the case of an elliptic cylinder [64]. In the dynamically coupled setting, the analog of this is a steadily translating configuration of a circular contour and a counter-rotating vortex pair, located symmetrically about the direction of motion. This was termed the moving Föppl equilibrium in [156] and is an equilibrium solution of system (3.44), for $N = 2$, under these symmetric conditions.

In this equilibrium, the point vortices lie on the following curves:

$$l_1^2 - R^2 = \pm 2 l_1 y_1, \tag{3.52}$$

where (x_1, y_1) are the coordinates of the point vortex relative to the center of the cylinder, Γ_1 is the strength of the vortex, R is the radius of the cylinder, and $l_1^2 = x_1^2 + y_1^2$. Since the left side of the equation is always positive, one chooses the $+$ sign if $y_1 > 0$ and the $-$ sign if $y_1 < 0$. The four branches are shown in Fig. 3.5 in non-dimensionalized coordinates $(x_1/R, y_1/R)$. Corresponding to the location of each pair, the cylinder forward speed V is given by

$$V = \frac{\Gamma_1}{4\pi} \frac{l_1^4}{y_1 (l_1^4 - R^4)}. \tag{3.53}$$

It is easily seen that for the same direction of cylinder motion the vortices in the pair must be opposite-signed.

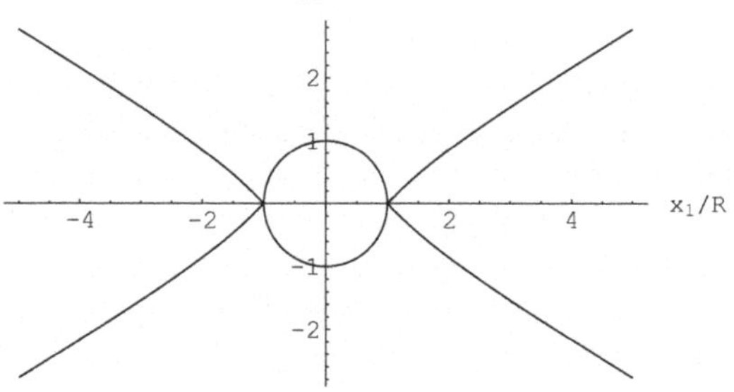

Fig. 3.5 The non-dimensionalized Föppl equilibrium curves. Reproduced from [151], with the permission of AIP Publishing

Each vortex in the translating pair has to satisfy both Eqs. (3.52) and (3.53). Since these equations are invariant under the transformation $x_1 \to -x_1$, it implies that for each pair position there is another position along the x-axis, reflected about the y-axis, which also satisfies the two equations (for the same V and Γ_1), as shown in Fig. 3.6. It is also important to note the signs of the vortices. In each configuration,

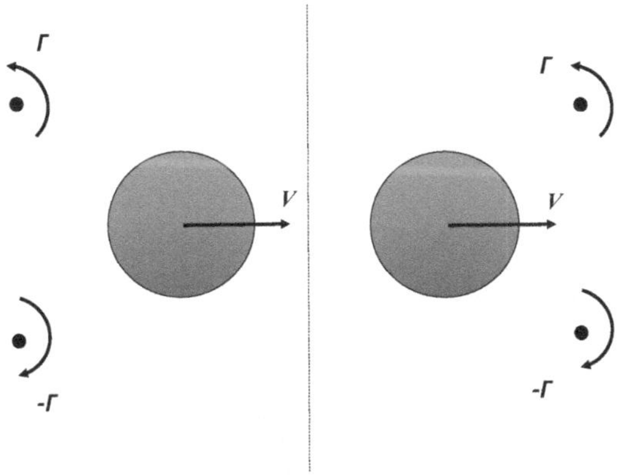

Fig. 3.6 Two configurations of the moving Föppl equilibrium, one in which the vortex pair is aft of the translating circle and one in which it is fore of the translating circle

keeping **V** the same, if the sign of Γ is switched, then the equilibrium is lost and the vortices move in the aftward direction. The cylinder maintains its direction, but its speed shows a temporary increase before asymptotically reaching a steady value. Referring to the discussion in Sect. 3.2.2 and Fig. 3.3, this latter configuration may be termed the swimming configuration since it results in increased forward momentum for the cylinder. The Föppl configuration however can be related to a different kind of momentum transfer scenario. One in which a pair of counter-rotating vortices of that sign approaches the cylinder at rest or moving slower than the vortices in the same direction. By the thumb rule proposed in Sect. 3.2.2 the flow linear momentum in the V-direction due to the vortices decreases during this approach. Conservation of the total linear momentum in that direction then results in the cylinder gaining forward speed. In this scenario the vortices therefore push the body in the same direction of traverse in contrast with a push in the opposite direction in the swimming scenario. In Chap. 5, viscous evolutions from such an initial configuration are considered. In the viscous setting there is obviously no translating equilibrium, but interestingly this kind of "hitching a ride from the vortices" scenario is still observed.

A detailed nonlinear stability analysis of this equilibrium using the energy-Casimir method, with respect to symmetry breaking and symmetry preserving perturbations, was also carried out in [156]. It may be recalled that the energy-Casimir method developed by Arnold [8], and later generalized to the energy-momentum method (see [12] and [109] for brief descriptions and references to important papers and examples), allows one to carry out a nonlinear stability analysis of an equilibrium point of a Hamiltonian system, i.e. a stability analysis with respect to finite perturbations as opposed to the infinitesimal perturbations used in traditional linearized stability analysis (for example, by Föppl in the classical equilibrium).

The dynamically coupled setting also introduces other important differences in the stability analysis—the main one being that the perturbations are not only to the positions of the vortices but also simultaneously to the coupled system linear momentum, and therefore to the cylinder velocity. In the classical equilibrium, on the other hand, the perturbations are typically to the positions of the vortices alone; see [28, 131, 162] for relatively more recent investigations of the classical Föppl equilibrium from both dynamics and control perspectives.

3.3.1 The Half-Space Model

The symmetric counter-rotating vortex pair configuration discussed above suggests a generalization, one in which the circle of radius R interacts with N_p *pairs* of such symmetrically located counter-rotating vortices ($N = 2N_p$), as shown in Fig. 3.7. In this model, throughout the system's evolution the reflection symmetry of the vortices about the **V**-direction is maintained and the circle's motion is constrained along the axis of symmetry. The phase space of this model can be identified with the phase space of the vortices in the top half alone and is therefore termed the "half-space" model. The equations of the point vortices are given by

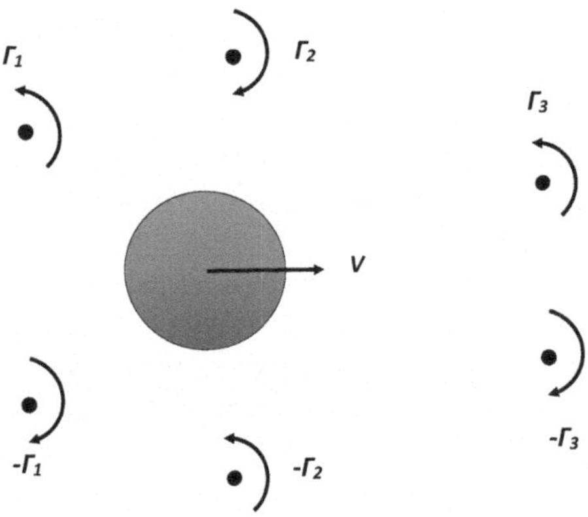

Fig. 3.7 An example of the "half-space" model showing a circle dynamically interacting with three pairs of counter-rotating point vortices located symmetrically about the **V**-direction

$$\frac{dx_j}{dt} = \frac{1}{2\Gamma_j}\frac{\partial H_{sym}}{\partial y_j},$$
$$\frac{dy_j}{dt} = -\frac{1}{2\Gamma_j}\frac{\partial H_{sym}}{\partial x_j}, \qquad j = 1, \cdots\cdots, N_p,$$

where

$$H_{sym} = \frac{1}{2c}\left(L - 2\sum_{j=1}^{N_p}\Gamma_j y_j + 2R^2\sum_{j=1}^{N_p}\Gamma_j\left(\frac{y_j}{x_j^2+y_j^2}\right)\right)^2 - \sum_{j=1}^{2N_p}\sum_{k>j}^{2N_p}\Gamma_k\Gamma_j G(\mathbf{l}_k;\mathbf{l}_j) - \frac{1}{2}\sum_{j=1}^{2N_p}\Gamma_j^2 g(\mathbf{l}_j;\mathbf{l}_j). \tag{3.54}$$

The conserved linear momentum variable $\mathbf{L}$ in this model is equivalent to a scalar and Eq. (3.45) becomes

$$L = cV + 2\sum_{j=1}^{N_p}\Gamma_j y_j - 2R^2\sum_{j=1}^{N_p}\Gamma_j\left(\frac{y_j}{x_j^2+y_j^2}\right), \tag{3.55}$$

where $c = 2\pi R^2$ and V is the cylinder speed.

The Case $N_p = 1$

The moving Föppl equilibrium can then also be viewed as a relative equilibrium of this model, when $N_p = 1$. There also exists a *normal line equilibrium* in which the two vortices are on the line perpendicular to the axis of symmetry and passing through the center of the circle. In this equilibrium, the vortices are located on the y-axis at $(0, y_1)$and $(0, -y_1)$. Assuming $\Gamma_1 > 0$ for the top vortex, the corresponding forward speed of the cylinder is given by the relation

$$-\left(V + \frac{\Gamma_1}{\pi}\left(\frac{y_1}{R^2} - \frac{1}{y_1}\right)\right)\left(1 + \frac{R^2}{y_1^2}\right) + \frac{\Gamma_1}{\pi R^2}y_1 - \frac{R^2\Gamma_1}{\pi y_1^3} + \Gamma_1\left(\frac{1}{2\pi}\frac{y_1}{y_1^2 - R^2} - \frac{1}{2\pi}\frac{(y_1^2+R^2)y_1}{y_1^4 + R^4 + 2R^2y_1^2} + \frac{1}{4\pi y_1}\right) = 0. \tag{3.56}$$

This model was studied in detail in [151] where among other things a bifurcation analysis of the case $N_p = 1$ was carried out with the bifurcation parameter defined as

$$\bar{L} = \frac{L}{\Gamma_1 R}.$$

Phase space trajectories, which are the trajectories of the top-half vortex, are shown for three typical values of $\bar{L}$ in Fig. 3.8. The bifurcation analysis reveals the following features:

1. Below the value of $\bar{L} \approx 1.744$, there are no equilibria. In such cases, the vortex pair passes over the cylinder.
2. The Föppl equilibria occur for values of $\bar{L} >\approx 1.744$. Moreover, for each value of $\bar{L}$ above this value, there are *two* locations of the vortex pairs (and correspondingly two values of the cylinder forward speed), one closer to the cylinder and

the other farther away. The closer one is stable and the farther one is unstable to symmetric perturbations.

3. The normal line equilibria occur for values of $\bar{L} >\approx 2.183$. Moreover, for each value of $\bar{L}$ above this value, there are *two* locations of the vortex pairs (and correspondingly two values of the cylinder forward speed), one closer to the cylinder and the other farther away. In this case, the closer one is unstable and the farther one is stable to symmetric perturbations.

The Case $N_p = 2$, Equal Strengths

Another moving equilibrium configuration is obtained when $N_p = 2$ and in which the vortex strengths are all equal in magnitude. The configuration is shown in Fig. 3.9. The positions of the vortices in this configuration are at four corners of a rectangle centered at the center of the circle. In other words, the positions are related as follows: if (x_1, y_1) and $(-x_1, y_1)$ locate the top-half pair, then $(x_1, -y_1)$ and $(-x_1, -y_1)$ locate the bottom-half pair, respectively.

The location of (x_1, y_1) is given by the curves in the top-half plane of Fig. 3.10. The equation governing these curves is a polynomial in x_1 and y_1 of 14th order. As in the case $N_p = 1$, for each (x_1, y_1) location there exists a unique value of the cylinder forward speed V. Here too, reversing the sign of the top-half vortices in Fig. 3.9 entails a reversing of the sign of $\mathbf{V}$; otherwise the equilibrium is lost. The method for obtaining V and other details, including results of a linear stability analysis and a bifurcation analysis, may be found in [151]. This equilibrium is also obtained in the classical setting of a stream of constant speed V at infinity approaching a fixed circular cylinder [43]. A linear stability analysis in this classical setting can be found in [119]. Here again, it should be emphasized that there is a difference in stability behavior of the equilibrium even though it occurs in both the dynamically coupled and classical settings.

Cases $N_p \geq 2$, Unequal Strengths

For $N_p \geq 2$, the question arises whether there exist equilibria in which the top-half vortices have unequal strengths and the associated configurations are not symmetric about the y-axis. The existence of such pairs in the classical setting was numerically demonstrated by Seath [149] and Weiss and Boasson [182] for the case when the vortices are in the wake region, i.e. aft of the cylinder, and in [119] for the case when the vortices are located on the normal line.

It is not hard to show that such equilibria must exist even in the dynamically coupled setting.

Proposition 3.5. *In the half-space model, for any N_p and for $L \neq 0$, there exists an infinite number of unequal strength translating equilibria in which the vortices are located on the normal line.*

Proof. First note that the expression for $G(\mathbf{l}_k, \mathbf{l}_j)$ given by (3.50) can be easily rewritten as

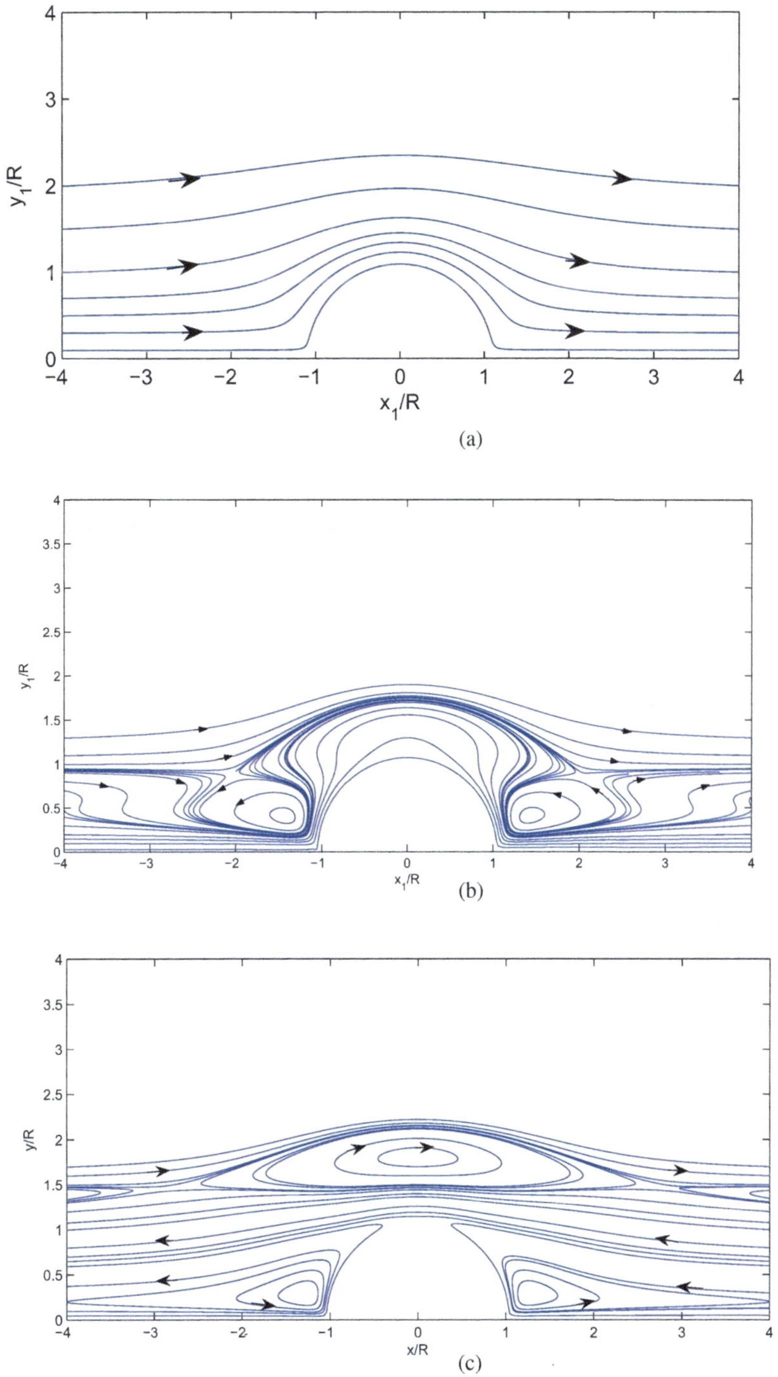

Fig. 3.8 Phase portraits for the case $N_p = 1$ for (**a**) $\bar{L} = 0$, (**b**) $\bar{L} = 2$, and (**c**) $\bar{L} = 3$ in the half-space model. Reproduced from [151], with the permission of AIP Publishing

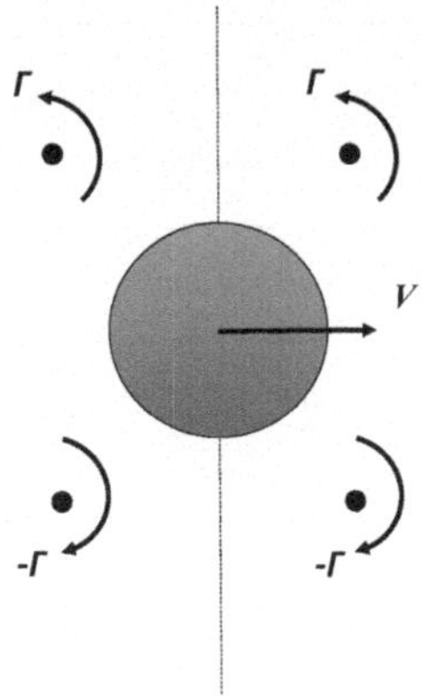

Fig. 3.9 A translating equilibrium with four point vortices

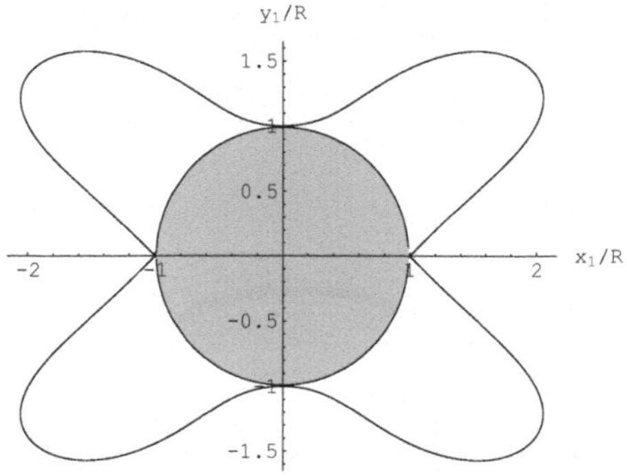

Fig. 3.10 The loci of equilibrium points in the same-signed symmetric equilibrium. Reproduced from [151], with the permission of AIP Publishing

$$G(\mathbf{l}_k,\mathbf{l}_j) = \frac{1}{4\pi}\left(\log(x_k^2+y_k^2)+\log(x_j^2+y_j^2)-\log\left(l_j^2 l_k^2+R^4-2R^2(x_jx_k+y_jy_k)\right)\right.$$
$$\left.+\log((x_k-x_j)^2+(y_k-y_j)^2)\right).$$

Using this and the expression for $g(\mathbf{l}_j,\mathbf{l}_j)$ given by (3.49), it is easy to see that the following is true:

$$\frac{\partial H_{sym}}{\partial x_j}\,|_{x_1=\cdots=x_{N_p}=0} = 0, \quad j = 1,\cdots,N_p$$

with H_{sym} given by (3.54).

It follows that normal line equilibria for any N_p are well-defined as solutions to the following system of N_p equations in the N_p variables $y_1, \cdots, y_{N_p}$.

$$\frac{dx_1}{dt}\Big|_{x_1=\cdots=x_{N_p}=0} = \frac{1}{2\Gamma_1}\frac{\partial H_{sym}}{\partial y_1}\Big|_{x_1=\cdots=x_{N_p}=0} = 0,$$
$$\cdots = \cdots = \cdots,$$
$$\frac{dx_{N_p}}{dt}\Big|_{x_1=\cdots=x_{N_p}=0} = \frac{1}{2\Gamma_{N_p}}\frac{\partial H_{sym}}{\partial y_{N_p}}\Big|_{x_1=\cdots=x_{N_p}=0} = 0.$$

Owing to the fact that the Hamiltonian vector field is linear in the Γ_js, and using (3.55), one obtains the following linear system:

$$\mathbf{A}_p \cdot \mathbf{V}_\Gamma = \frac{L}{R}\mathbf{B}, \tag{3.57}$$

where $\mathbf{A}_p$ is an $N_p \times N_p$ matrix whose elements are functions of $\bar{y}_1, \cdots, \bar{y}_{N_p}$ ($\bar{y}_j := y_j/R$), $\mathbf{V}_\Gamma = (\Gamma_1, \cdots, \Gamma_{N_p})$, and

$$\mathbf{B} = \begin{pmatrix} 1+\frac{1}{\bar{y_1}^2} \\ \cdots \\ \cdots \\ 1+\frac{1}{\bar{y_{N_p}}^2} \end{pmatrix}. \tag{3.58}$$

Alternatively, one obtains a different linear system in the Γ_js by replacing V with L using (3.55). Either way, one has N_p equations in $2N_p+1$ unknown parameters: $\Gamma_1, \cdots, \Gamma_{N_p}, \bar{y}_1, \cdots, \bar{y}_{N_p}$, and L/R. Therefore, N_p+1 of these parameters can be chosen arbitrarily. Note that (3.57) may also be viewed as a system of N_p high-order polynomial equations in $\bar{y}_1, \cdots, \bar{y}_{N_p}$, but from a computational point of view it is obviously much easier to view it as a linear system in the Γ_js.

To obtain an equilibrium configuration, first assign values to the entries of the vector $\bar{y}_1, \cdots, \bar{y}_{N_p}$. These values must satisfy the following conditions: no two of them are equal, each is > 1 and $\mathrm{Det}\,\mathbf{A}_p \neq 0$. Next, invert the system (3.57) to relate the vector $\mathbf{V}_\Gamma$ and L/R. Note that the equation $\mathrm{Det}\,\mathbf{A}_p = 0$ determines, at the most, a surface of dimension $N_p - 1$ in $\mathbb{R}^{N_p} \backslash (\triangle \cup B^{N_p})$ and therefore, there should exist an infinite number of such equilibrium configurations. For $L/R = 0$, the equilibrium locations are defined by the real solutions of $\mathrm{Det}\,\mathbf{A}_p = 0$, if they exist. Note that one can also recast (3.57) as a homogeneous system by extending the vector of unknowns to $\mathbf{V}_{\Gamma,L} := (L, \Gamma_1, \cdots, \Gamma_{N_p})$, and solve using standard linear algebra techniques.

Example: Normal Line Equilibria for the Case $N_p = 2$

As an example of the above proposition consider the case $N_p = 2$. The Hamiltonian function and vector field are presented in Appendix C.

Substituting $x_1 = x_2 = 0$ in the vector field equations of Appendix C, it is easily seen that $dy_1/dt = dy_2/dt = 0$. And so the conditions for equilibria are given by the equations

$$\left(-\frac{1}{c}-\frac{R^2}{cy_1^2}\right)\left(L-2\Gamma_2 y_2+\frac{2R^2\Gamma_2}{y_2}\right)+\frac{2\Gamma_1}{c}y_1-\frac{2R^4\Gamma_1}{c}\frac{1}{y_1^3}$$
$$+\Gamma_1\left(\frac{1}{2\pi}\frac{y_1}{y_1^2-R^2}-\frac{1}{2\pi}\frac{y_1}{y_1^2+R^2}+\frac{1}{4\pi y_1}\right)+\frac{\Gamma_2}{4\pi}4y_2\left(\frac{R^2}{y_1^2y_2^2-R^4}-\frac{1}{y_1^2-y_2^2}\right)=0 \tag{3.59}$$

$$\left(-\frac{1}{c}-\frac{R^2}{cy_2^2}\right)\left(L-2\Gamma_1 y_1+\frac{2R^2\Gamma_1}{y_1}\right)+\frac{2\Gamma_2}{c}y_2-\frac{2R^4\Gamma_2}{c}\frac{1}{y_2^3}$$
$$+\Gamma_2\left(\frac{1}{2\pi}\frac{y_2}{y_2^2-R^2}-\frac{1}{2\pi}\frac{y_2}{y_2^2+R^2}+\frac{1}{4\pi y_2}\right)+\frac{\Gamma_1}{4\pi}4y_1\left(\frac{R^2}{y_1^2y_2^2-R^4}+\frac{1}{y_1^2-y_2^2}\right)=0, \tag{3.60}$$

which leads to the following system of linear equations in $\bar{y}_1$ and $\bar{y}_2$,

$$\begin{pmatrix} f(\bar{y}_1) & g(\bar{y}_1,\bar{y}_2) \\ g(\bar{y}_2,\bar{y}_1) & f(\bar{y}_2) \end{pmatrix}\cdot\begin{pmatrix}\Gamma_1\\ \Gamma_2\end{pmatrix}=\frac{L}{R}\begin{pmatrix}1+\frac{1}{\bar{y}_1^2}\\ 1+\frac{1}{\bar{y}_2^2}\end{pmatrix}, \tag{3.61}$$

where

$$f(\bar{y}) := 2\bar{y}-\frac{2}{\bar{y}^3}+\left(\frac{\bar{y}}{\bar{y}^2-1}-\frac{\bar{y}}{\bar{y}^2+1}+\frac{1}{2\bar{y}}\right),$$
$$g(\bar{y}_1,\bar{y}_2) := \left(-1-\frac{1}{\bar{y}_1^2}\right)\left(-2\bar{y}_2+\frac{2}{\bar{y}_2}\right)+2\bar{y}_2\left(\frac{1}{\bar{y}_1^2\bar{y}_2^2-1}-\frac{1}{\bar{y}_1^2-\bar{y}_2^2}\right).$$

The vortex strengths are then given by

$$\begin{pmatrix}\Gamma_1\\ \Gamma_2\end{pmatrix}=\frac{L}{R}\mathbf{A}_p^{-1}\cdot\begin{pmatrix}1+\frac{1}{\bar{y}_1^2}\\ 1+\frac{1}{\bar{y}_2^2}\end{pmatrix} \tag{3.62}$$

and the cylinder speed is obtained from (3.55) as

$$2\pi RV = \frac{L}{R}+\sum_{j=1}^{2}2\Gamma_j\left(\frac{1}{\bar{y}_j}-\bar{y}_j\right).$$

For some sample choices of $\bar{y}_1$ and $\bar{y}_2$, the values of Γ_1 and Γ_2 and $2\pi RV$ are shown in Table 3.1. It may be noted that in all three cases, the vortex closer to the cylinder has clockwise circulation and the vortex farther away has counter-clockwise circulation. A schematic sketch of the equilibrium configuration is shown in Fig. 3.11.

Table 3.1 Table of sample values in the normal line equilibrium for $N_p = 2$ and with $R = 1$. For the first two rows, $L = 1$, for the last row $L = 2$

$\bar{y}_1$	$\bar{y}_2$	Γ_1	Γ_2	$2\pi RV$
2	3	−0.243	0.284	0.211
5	10	−0.078	0.087	0.144
2	3	−0.049	0.57	0.423

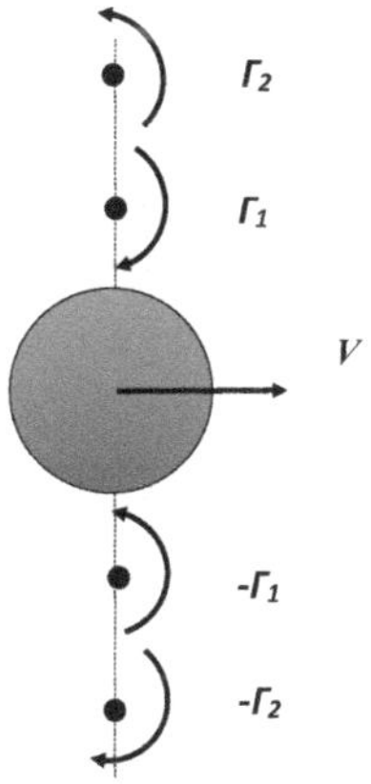

Fig. 3.11 A normal line translating equilibrium configuration with two pairs of symmetric counter-rotating vortices, $\Gamma_1 \neq \Gamma_2$ and are opposite-signed

Wake Equilibria for $N_p = 2$

From the above investigations, it should be clear that the existence of translating equilibria in which the point vortices do not lie on the y-axis for any N_p is also governed by a linear system similar to (3.57), but the matrix $\mathbf{A}_p$ is of size $2N_p \times N_p$ with entries that in general are functions of $(x_1, \cdots, x_N, y_1, \cdots, y_N)$. Consider the case $N_p = 2$. The configuration in Fig. 3.9 is an example of this case, but the vortices have an additional reflective symmetry with respect to the y-axis. In this section, additional equilibrium configurations that do not posses this y-axis symmetry are discussed. In particular, in these configurations the two vortex pairs lie on the same "side" of the cylinder—in fact in the wake region—and are hence referred to as wake equilibria. The numerical recipe for finding these equilibria is as follows. From the equations of motion of the two-pair system presented in Appendix C, one again obtains the equilibrium configurations as solutions to an algebraic system of the form (3.57), but $\mathbf{A}_p$ is now of size 4×2 and the entries of both $\mathbf{A}_p$ and the vector $\mathbf{B}$ are functions of $\bar{x}_1, \bar{y}_1, \bar{x}_2$, and $\bar{y}_2$. Standard linear algebra then states that for a unique solution $\mathbf{V}_\Gamma$ to exist, $\text{Rank}\,\mathbf{A}_p = \text{Rank}\,\tilde{\mathbf{A}}_p = 2$, where $\tilde{\mathbf{A}}_p := (\mathbf{A}_p, \mathbf{b})$ is the 4×3 augmented matrix. Denoting by $\mathbf{D}_i, i = 1, \cdots, 4$ the four 3×3 submatrices of $\tilde{\mathbf{A}}_p$, it follows that the necessary and sufficient condition for these equilibria to exist is that at the equilibrium locations of the point vortices,

$$\text{Det } \mathbf{D}_i = 0, \quad i = 1, \cdots, 4, \tag{3.63}$$

and at least one 2×2 submatrix of $\mathbf{A}_p$ and $\tilde{\mathbf{A}}_p$ has non-zero determinant. The equilibrium locations of the point vortices therefore lie in the intersection set of real solutions to (3.63), if non-empty. Each of these equations defines, generically, a three-dimensional manifold in $\mathbb{R}^4$, making the task of finding the intersection set challenging. It is computationally easier to arbitrarily assign the location $(\bar{x}_1, \bar{y}_1)$ of one vortex and seek the intersection set, now the set of $(\bar{x}_1, \bar{y}_1)$, as the common intersection point(s) of four curves in the plane. Two examples of wake equilibria obtained by this numerical method, for both of which $L = R = 1$:

$$\begin{aligned} (a) \quad \{x_1, y_1, x_2, y_2\} &= \{-2, 2, -1.817, 0.642\} \\ \{\Gamma_1, \Gamma_2\} &= \{0.306, -0.251\}, \\ 2\pi RV &= 0.21. \end{aligned}$$

$$\begin{aligned} (b) \quad \{x_1, y_1, x_2, y_2\} &= \{-2, 1, -5.131, 1.977\} \\ \{\Gamma_1, \Gamma_2\} &= \{-0.005, 0.247\}, \\ 2\pi RV &= 0.226. \end{aligned}$$

In these examples again, one may note that the pairs have opposite circulations. It is conjectured that there exists an infinite number of such wake equilibria.

The rigid circle case offers the simplest geometry, and due to the physical domain being planar, it also allows the use of conformal map techniques to study non-circular rigid contours. In particular, the case of an elliptic contour dynamically interacting with point vortices has been studied using such techniques. Topics studied include analogues of the moving Föppl and normal line equilibria [64, 80] and non-integrable and chaotic behavior [25, 140].

Chapter 4
Dynamically Coupled Rigid Body+Vortex Rings in $\mathbb{R}^3$

The dynamically coupled planar models of the previous chapter are now extended to $\mathbb{R}^3$, with point vortices replaced by vortex rings. The streamfunction is replaced by the vector potential (2.8). As in the planar models, the problem of a stationary body is first considered before moving on to the dynamically coupled problem.

Before discussing these models, it is useful to keep the following fact about compact vorticity in $\mathbb{R}^3$ in mind.

Total Vorticity

In the previous chapter, cases were distinguished on the basis of whether the quantity Γ, termed as the total vortex strength of the point vortices, is zero or not. Referring to (2.10) it is clear that this sum is equal to the *total vorticity* integral,

$$\Gamma \mathbf{k} = \int_D \omega\, \mu,$$

for a distribution of point vortices.

For vortex rings in $\mathbb{R}^3$, however, it turns out that the two notions are different. The total vortex strength Γ is still defined and can take on different values, but the total vorticity is equal to zero in all cases. The latter is a consequence of the following important result:

A smooth distribution of vorticity in $\mathbb{R}^3$ that decays faster than $1/r^3$, as $r \to \infty$, has zero total vorticity

The result simply follows from the following identity for divergence-free vector fields, obtained by applying the divergence theorem in $\mathbb{R}^3$:

$$\int_U \omega\, \mu = -\int_{\partial U} \mathbf{r}(\omega \cdot \mathbf{n})\, \nu, \quad U \subset \mathbb{R}^3, \tag{4.1}$$

B. N. Shashikanth, *Dynamically Coupled Rigid Body-Fluid Flow Systems*,
https://doi.org/10.1007/978-3-030-82646-8_4

and observing that for the assumed decay rates the right side goes to zero as $U \to \mathbb{R}^3$. Note that in the presence of boundaries, one adds the requirement that the vorticity is either zero, or has a zero normal component, everywhere on the boundary.

To apply the above identity to a compact but singular vorticity distribution such as rings, one constructs tubular neighborhoods of the rings as mentioned at the end of Sect. 3.1 (and also used extensively in the next section). By taking the external boundary of U to be large enough to enclose all the rings, it is easy to show that the result holds for rings as well and regardless of the total vortex strength of the rings:

$$\int_D \omega\, \mu = 0 \quad \text{for any } \Gamma. \tag{4.2}$$

Equation (4.1) cannot be applied to domains in $\mathbb{R}^2$ since the vorticity field is perpendicular to the plane. The divergence theorem in $\mathbb{R}^2$ applies to vector fields lying in the plane. The identity can still be applied if one views point vortices as linear parallel infinitely long vortex filaments in $\mathbb{R}^3$, but this fails to provide any result.

Example.

Consider a circular vortex ring of strength Γ with a core of circular cross-section. In cylindrical coordinates (r, θ, z) assume the vorticity field is $\omega(\theta) = \mid \omega \mid (\theta)\hat{e}_\theta$ with $\mid \omega \mid$=constant everywhere. It follows that, regardless of the value of Γ,

$$\int_{\mathbb{R}^3} \omega\, \mu = \int_{z_1}^{z_2} \int_{R_1}^{R_2} \int_0^{2\pi} \omega(\theta)\, r dr d\theta dz = \int_0^{2\pi} \omega(\theta)\, d\theta \int_{z_1}^{z_2} \int_{R_1}^{R_2} r dr dz = 0$$

since $\int_0^{2\pi} \omega(\theta)\, d\theta = \mid \omega \mid \int_0^{2\pi} \hat{e}_\theta\, d\theta = 0$.

4.1 *N* Vortex Rings and a Single Stationary Rigid Boundary

As in the planar problem, it is instructive to start with the problem in which the rings are in an externally unbounded domain $D \subset \mathbb{R}^3$ with a single internal rigid boundary, the surface of a smooth and compact body B that is stationary. This could be considered a three-dimensional analog of C. C. Lin's problem and, in similar manner, serves as an intermediate model that helps in understanding the dynamically coupled model more easily.

Following the discussion in Sect. 3.1, consider first the vector potential due to the jth ring of strength Γ_j in unbounded flow, with s_j the arc-length parameter:

$$\mathbf{A}_j(p) = \frac{\Gamma_j}{4\pi} \oint_{C_j} \frac{\mathbf{t}_j(s_j) ds_j}{\mid \mathbf{r}(p) - \mathbf{r}_j(s_j) \mid}, \quad p \in \mathbb{R}^3. \tag{4.3}$$

Recall that the vector potential is obtained as a solution to Eq. (2.9). The velocity field due to the jth ring in unbounded flow is obtained as $\nabla \times \mathbf{A}_j$ or directly from (2.24) as

$$\mathbf{v}_j(p) = \frac{\Gamma_j}{4\pi} \oint_{C_j} \frac{\mathbf{t}_j(s_j) \times (\mathbf{r}(p) - \mathbf{r}_j(s_j))}{\left|\mathbf{r}(p) - \mathbf{r}_j(s_j)\right|^3} ds_j, \quad p \in \mathbb{R}^3. \tag{4.4}$$

From (4.3) and (4.4), the far-field fields must decay at least as rapidly as $1/\mid \mathbf{r}(p) \mid$ and $1/\mid \mathbf{r}(p) \mid^2$, respectively. But because of (4.2), the actual decay rate is even faster [165]:

$$\mathbf{A}_j(p) \sim \frac{1}{\mid \mathbf{r}(p) \mid^2}, \quad \mathbf{v}_j(p) \sim \frac{1}{\mid \mathbf{r}(p) \mid^3}, \quad \mid \mathbf{r}(p) \mid \to \infty. \tag{4.5}$$

To construct an image field for a ring of unit strength, requires information about both its location with respect to the boundaries and its shape, since both these parameters determine the vorticity distribution relative to the boundaries. Since the rings are viewed as images of maps $C_j : S^1 \to \mathbb{R}^3$, which are independent of the strength of the ring, let the single vector potential analogous to the function $\psi^u_{I,j}(\mathbf{r})$ of Sect. 3.1 be denoted by $\mathbf{A}^u_{I,j}(p)$, the vector potential of the image field due to a ring of *unit* strength located at C_j. Let $\mathbf{v}^u_I = \nabla \times \mathbf{A}^u_I$ denote the corresponding velocity field.

The vector potential and the velocity field due to N rings including their images can then be written as:

$$\mathbf{A}_\omega(p) = \sum \left(\mathbf{A}_j(p) + \Gamma_j \mathbf{A}^u_{I,j}(p) \right), \tag{4.6}$$

$$\mathbf{v}_\omega(p) = \sum \left(\mathbf{v}_k(p) + \Gamma_k \mathbf{v}^u_{I,k}(p) \right). \tag{4.7}$$

In applying (2.7) to compute the kinetic energy, a punctured domain as discussed in Sect. 3.1 will be used, since improper integrals are again involved. But as with point vortices most of these terms can be computed more easily using an unpunctured domain and substituting the vorticity distribution (2.11) directly in (2.7) and making use of properties of Dirac delta distributions. The punctured domain is necessary only for computing the self-interaction terms. Nevertheless, to be consistent and since it is the mathematically more correct way, it is shown below how to compute all the terms using such a domain. The construction of this domain is explained in [157] and is recalled here.

Enclose each ring within a solid torus of constant circular cross-section. Such a torus can be easily constructed: At each point of the ring draw a circle of radius ε lying in the normal plane of the ring at that point. The union of these discs is the solid torus.[1] Denote by T^ε_j the solid torus of the jth ring and apply (2.7) to the domain

---

[1] Since it is assumed that the curve has finite curvature at all points, an ε small enough so that the circles do not intersect can always be chosen.

$$D^\varepsilon := D \backslash (T_1^\varepsilon \cup T_2^\varepsilon \cdots T_N^\varepsilon),$$
$$\Rightarrow \partial D^\varepsilon := \partial B \cup \partial T_1^\varepsilon \cup \partial T_2^\varepsilon \cdots \partial T_N^\varepsilon,$$

where ∂B denotes the union of all the rigid boundaries. This gives

$$\int_D \langle \mathbf{v}, \mathbf{v} \rangle_{\mathbb{R}^3} \mu = \lim_{\varepsilon \to 0} \int_{D^\varepsilon} \langle \mathbf{v}, \mathbf{v} \rangle_{\mathbb{R}^3} \mu = - \int_{\partial B} \mathbf{A}_\omega \times \mathbf{v}_\omega \cdot \mathbf{n}\, \nu - \lim_{\varepsilon \to 0} \int_{\partial D^\varepsilon} \mathbf{A}_\omega \times \mathbf{v}_\omega \cdot \mathbf{n}\, \nu$$
$$- \lim_{R \to \infty} \int_{S_R} \mathbf{A}_\omega \times \mathbf{v}_\omega \cdot \mathbf{n}\, \nu$$

recalling again the convention for the unit normal field on fluid boundaries, and where S_R denotes a sphere of radius R enclosing all the rings and B. The vector potential field on ∂B, in general, is neither normal or tangent to the surface. However, similar to the streamfunction, the vector potential has a "gauge freedom." It is defined only up to an additive gradient field. It is shown in [157] (Proposition 4.1 *op. cit.*), and implied in the earlier work of Hirasaki and Hellums [65], how to adjust the vector potential on ∂B by the addition of a gradient field such that the adjusted field is normal to the surface. This property of a vector potential field is, in fact, quite general and is not specific to the field associated with a vortex ring. The first surface integral on the right can in this way be made to vanish.

Substituting (4.6) and (4.7) into the second integral on the right and considering only the integral around the torus of the jth ring,

$$\begin{aligned}
\int_{\partial T_j} \mathbf{A}_\omega \times \mathbf{v}_\omega \cdot \mathbf{n}\, \nu = \sum_{i,k} \int_{\partial T_j} \Big(\mathbf{A}_i(p) \times \mathbf{v}_k(p) \Big) \cdot \mathbf{n}\, \nu \\
+ \sum_{i,k} \Gamma_i \int_{\partial T_j} \Big(\mathbf{A}^u_{I,i}(p) \times \mathbf{v}_k(p) \Big) \cdot \mathbf{n}\, \nu \\
+ \sum_{i,k} \Gamma_k \int_{\partial T_j} \Big(\mathbf{A}_i(p) \times \mathbf{v}^u_{I,k}(p) \Big) \cdot \mathbf{n}\, \nu \\
+ \sum_{i,k} \Gamma_i \Gamma_k \int_{\partial T_j} \Big(\mathbf{A}^u_{I,i}(p) \times \mathbf{v}^u_{I,k}(p) \Big) \cdot \mathbf{n}\, \nu .
\end{aligned} \tag{4.8}$$

Equip the surface of the torus with coordinates (s_j, θ), where (r_j, θ) are polar coordinates of the plane which contains the circle of radius ε mentioned above. The area form on the torus surface can then be represented as $\nu = \varepsilon d\theta ds_j$, and each integral can be evaluated as

$$\int_{\partial T_j} (\cdots) \cdot \mathbf{n}\, \nu \equiv \oint \int_0^{2\pi} (\cdots) \cdot \mathbf{n}\, \varepsilon d\theta ds_j$$

Consider now the last set of integrals on the right side of (4.8). Since the $\mathbf{A}_I$ and $\mathbf{v}_I$ are vector fields whose components are harmonic functions in D, and in particular $O(1)$ on each ring, it is clear that for each i (or k)

$$\mathbf{A}^u_{I,i}(p) \equiv \mathbf{A}^u_{I,i}(s_j,\theta) = \mathbf{A}^u_{I,i}(s_j) + O(\varepsilon), \quad p \in \partial T_j,$$
$$\mathbf{v}^u_{I,k}(p) \equiv \mathbf{v}^u_{I,k}(s_j,\theta) = \mathbf{v}^u_{I,k}(s_j) + O(\varepsilon).$$

And so for each i,k (including $i = j$, $k = j$),

$$\int_{\partial T_j} \left(\mathbf{A}^u_{I,i}(p) \times \mathbf{v}^u_{I,k}(p)\right) \cdot \mathbf{n}\, \nu$$
$$= \oint \left(\mathbf{A}^u_{I,i}(s_j) \times \mathbf{v}^u_{I,k}(s_j)\right) \cdot \int_0^{2\pi} \mathbf{n}\, \varepsilon d\theta ds_j + \oint \int_0^{2\pi} O(\varepsilon) \cdot \mathbf{n}\, \varepsilon d\theta ds_j$$
$$\Rightarrow \lim_{\varepsilon \to 0} \int_{\partial T_j} \left(\mathbf{A}^u_{I,i}(p) \times \mathbf{v}^u_{I,k}(p)\right) \cdot \mathbf{n}\, \nu = 0 \quad \forall i,k.$$

Consider next the third set of integrals on the right side of (4.8). For $i \neq j, k \neq j$, one has, just as for the image fields,

$$\begin{aligned} \mathbf{A}_i(p) &= \mathbf{A}_i(s_j) + O(\varepsilon), \\ \mathbf{v}_k(p) &= \mathbf{v}_k(s_j) + O(\varepsilon). \end{aligned} \tag{4.9}$$

Using the first series, it follows that for any k,

$$\lim_{\varepsilon \to 0} \int_{\partial T_j} \left(\mathbf{A}_i(p) \times \mathbf{v}^u_{I,k}(p)\right) \cdot \mathbf{n}\, \nu = 0, \quad i \neq j.$$

For $i = j$, using the leading order behavior expressions in Appendix B, obtain

$$\int_{\partial T_j} \left(\mathbf{A}_j(p) \times \mathbf{v}^u_{I,k}(p)\right) \cdot \mathbf{n}\, \nu \tag{4.10}$$
$$= \oint \int_0^{2\pi} \left(\left(\frac{\Gamma_j \mathbf{t}_j(s_f)}{4\pi} K(\varepsilon; l) + O(1) \right) \times \left(\mathbf{v}^u_{I,k}(s_j) + O(\varepsilon) \right) \right) \cdot \mathbf{n}\, \varepsilon d\theta ds_j, \tag{4.11}$$

where

$$K(\varepsilon; l) := \left\{ \log\left(l + \sqrt{l^2 + \varepsilon^2}\right) - \log\left(-l + \sqrt{l^2 + \varepsilon^2}\right) \right\}$$

and the second term in $\mathbf{A}_j$ represents any possible contribution from $\mathbf{A}_{j,non-local}$ (see discussion at the end of Appendix B). Since $K(\varepsilon; l) = O(\log \varepsilon)$, it follows that

$$\lim_{\varepsilon \to 0} \int_{\partial T_j} \left(\mathbf{A}_j(p) \times \mathbf{v}^u_{I,k}(p)\right) \cdot \mathbf{n}\, \nu = 0.$$
$$\Rightarrow \lim_{\varepsilon \to 0} \int_{\partial T_j} \left(\mathbf{A}_i(p) \times \mathbf{v}^u_{I,k}(p)\right) \cdot \mathbf{n}\, \nu = 0, \quad \forall j,k.$$

Consider next the second set of integrals on the right side of (4.8). For $k \neq j$, any i, one obtains

$$\lim_{\varepsilon \to 0} \int_{\partial T_j} \left(\mathbf{A}^u_{I,i}(p) \times \mathbf{v}_k(p) \right) \cdot \mathbf{n}\, \nu = 0.$$

And for $k = j$, any i, using the expressions in Appendix B,

$$\begin{aligned} &\int_{\partial T_j} \left(\mathbf{A}^u_{I,i}(p) \times \mathbf{v}_j(p) \right) \cdot \mathbf{n}\, \nu \\ &= \oint \int_0^{2\pi} \left(\left(\mathbf{A}^u_{I,i}(s_j) + O(\varepsilon) \right) \right. \qquad (4.12) \\ &\qquad \left. \times \left(\frac{\Gamma_j}{4\pi} \left[\frac{2l}{\sqrt{\varepsilon^2 + l^2}} \left(\frac{\cos\theta}{\varepsilon} \mathbf{j} - \frac{\sin\theta}{\varepsilon} \mathbf{i} \right) + O(\log\varepsilon) \right) \right) \right) \cdot \mathbf{n}\, \varepsilon d\theta ds_j, \\ &= \oint \int_0^{2\pi} \left(\left(\mathbf{A}^u_{I,i}(s_j) + O(\varepsilon) \right) \times \left(\frac{\Gamma_j}{4\pi} \left[\frac{2l}{\sqrt{\varepsilon^2 + l^2}} \frac{\mathbf{e}_\theta}{\varepsilon} + O(\log\varepsilon) \right) \right) \right) \cdot \mathbf{n}\, \varepsilon d\theta ds_j, \\ &\qquad (4.13) \end{aligned}$$

where $\mathbf{e}_\theta$ is the unit tangent vector field on the circle taken to be positive in the counter-clockwise direction. The orientation of ∂T_j is defined by $\mathbf{n} \times \mathbf{e}_\theta = \mathbf{t}_j$. As $\varepsilon \to 0$, the vector field $\mathbf{e}_\theta \times \mathbf{n}$ converges uniformly to $-\mathbf{t}_j$ at p_{f}. It follows that

$$\lim_{\varepsilon \to 0} \int_{\partial T_j} \left(\mathbf{A}^u_{I,i}(p) \times \mathbf{v}_j(p) \right) \cdot \mathbf{n}\, \nu = -\Gamma_j \oint_{C_j} \mathbf{A}_{I,i}(s_j) \cdot \mathbf{t}_j(s_j) ds_j.$$

Finally, consider the first set of integrals on the right side of (4.8). There are four cases here to consider. For $i, k \neq j$, it is easy to see from (4.9) that the integral has the same behavior as the fourth set of integrals evaluated previously. Therefore,

$$\lim_{\varepsilon \to 0} \int_{\partial T_j} \left(\mathbf{A}_i(p) \times \mathbf{v}_k(p) \right) \cdot \mathbf{n}\, \nu = 0, \quad i, k \neq j.$$

For $i = j, k \neq j$, the behavior is similar to (4.11) and one obtains again

$$\lim_{\varepsilon \to 0} \int_{\partial T_j} \left(\mathbf{A}_j(p) \times \mathbf{v}_k(p) \right) \cdot \mathbf{n}\, \nu = 0, \quad k \neq j.$$

For $i \neq j, k = j$, the behavior is similar to (4.13) and one obtains again

$$\lim_{\varepsilon \to 0} \int_{\partial T_j} \left(\mathbf{A}_i(p) \times \mathbf{v}_j(p) \right) \cdot \mathbf{n}\, \nu = -\Gamma_j \oint_{C_j} \mathbf{A}_i(s_j) \cdot \mathbf{t}_j(s_j) ds_j.$$

For $i = k = j$, the expressions in Appendix B give

$$\int_{\partial T_j} \Big(\mathbf{A}_j(p) \times \mathbf{v}_j(p)\Big) \cdot \mathbf{n}\, \nu$$
$$= \oint \int_0^{2\pi} \left(\left(\frac{\Gamma_j \mathbf{t}_j(s_f)}{4\pi} K(\varepsilon; l) + O(1) \right) \times \left(\frac{\Gamma_j}{4\pi} \left[\frac{2l}{\sqrt{\varepsilon^2 + l^2}} \frac{\mathbf{e}_\theta}{\varepsilon} + O(\log \varepsilon) \right) \right) \cdot \mathbf{n}\, \varepsilon d\theta ds_j,$$
$$\Rightarrow \lim_{\varepsilon \to 0} \int_{\partial T_j} \Big(\mathbf{A}_j(p) \times \mathbf{v}_j(p)\Big) \cdot \mathbf{n}\, \nu = \lim_{\varepsilon \to 0} \frac{\Gamma_j^2}{2\pi} \log \varepsilon \oint ds_j,$$

using the fact that $\lim_{\varepsilon \to 0} K(\varepsilon, l) := -2 \log \varepsilon + O(1)$. The blow-up in the self-induced fields results in this divergent contribution to the kinetic energy.

To summarize, the contribution to the kinetic energy of N rings due to the jth ring alone is of three kinds:

* The sum
$$\sum_i \Gamma_i \Gamma_j \oint_{C_j} \mathbf{A}^u_{I,i}(s_j) \cdot \mathbf{t}_j(s_j) ds_j,$$
representing the mutual interaction of the field of the jth ring with the image field of each ring, including its own.

* The sum
$$\sum_{\substack{i \\ i \neq j}} \Gamma_j \oint_{C_j} \mathbf{A}_i(s_j) \cdot \mathbf{t}_j(s_j) ds_j,$$
representing the mutual interaction of the field of the jth ring with the field of every other ring.

* The term
$$\lim_{\varepsilon \to 0} -\frac{\Gamma_j^2}{2\pi} \log \varepsilon \oint ds_j, \tag{4.14}$$
appropriately regularized, representing the self-interaction of the field of the jth ring with itself.

Finally, note that the decay rates of each $\mathbf{A}_j$ and each $\mathbf{v}_j$, as also the image fields, ensure that

$$\lim_{R \to \infty} \int_{S_R} \mathbf{A}_\omega \times \mathbf{v}_\omega \cdot \mathbf{n}\, \nu = 0.$$

Gathering all terms above computed for the jth ring and summing over the N rings provide the expression for the kinetic energy for N vortex rings interacting in the presence of a single stationary rigid body.

Referring to Sect. 2.2.2 (the lines above Eq. (2.31)), the phase space of the N rings in the presence of the stationary body can still be denoted by $P_{\rm f}^N$ but with each ring now identified with the image of the map $C_j : S^1 \to \mathbb{R}^3 \backslash B$. The evolution equations of the N rings have an additional term on the right, when compared with (2.34), a consequence of the image velocity fields:

$$\frac{\partial C_j}{\partial t} = \sum_{i \neq j} \frac{\Gamma_i}{8\pi} \oint_{C_i} \left(\frac{\mathbf{t}_i(s_i) \times \mathbf{R}_{ji}}{r_{ji}^3} \right)_{n_j} ds_i + \frac{1}{\Gamma_j} \left(\frac{\delta H_{\rm SI,reg}^j}{\delta C_j} \right)^\sharp \times \mathbf{t}_j + \Gamma_j (\mathbf{v}_I)_{n_j}, \qquad j = 1, \ldots, N \tag{4.15}$$

Recall that the subscript n_j denotes the component of the velocity projected onto the normal plane of the curve.

Proposition 4.1. *The system of N vortex rings dynamically interacting in the presence of a single stationary rigid body (compact and smooth) is Hamiltonian with respect to the Marsden–Weinstein Poisson structure (2.32) on* $P_{\rm f}^N$ *and Hamiltonian function* $H : P_{\rm f}^N \to \mathbb{R}$ *defined by*

$$\begin{aligned} &2H(C_1, \cdots, C_N) \\ &:= \sum_{\substack{i,j \\ i \neq j}} \Gamma_j \oint_{C_j} \mathbf{A}_i(s_j) \cdot \mathbf{t}_j(s_j) ds_j + \sum_{i,j} \Gamma_i \Gamma_j \oint_{C_j} \mathbf{A}_{I,i}^u(s_j) \cdot \mathbf{t}_j(s_j) ds_j + 2 \sum H_{\rm SI,reg}^j, \end{aligned} \tag{4.16}$$

where $H_{\rm SI,reg}^j$ *is the self-interaction kinetic energy, (4.14), appropriately regularized.*

Proof. The proof consists of showing that the Hamiltonian vector field of H above, $X_H : P_{\rm f}^N \to TP_{\rm f}^N$, with respect to (2.32) can be identified with the right side of (4.15).

The proof is made easier by first re-writing H above in a slightly different manner. Referring to (4.6), let

$$\mathbf{A}_{\omega,j}(p) := \mathbf{A}_j(p) + \Gamma_j \mathbf{A}_{I,j}(p), \quad p \in D. \tag{4.17}$$

It turns out that $\mathbf{A}_{\omega,j}$ has several interesting properties, as shown in [157], one of which is (Proposition 4.3, *op. cit.*): for any two vortex rings, C_i, C_j, in the presence of a body, the following reciprocity relation holds:

$$\Gamma_j \oint_{C_j} \mathbf{A}_{\omega,i}(s_j) \cdot \mathbf{t}_j(s_j) ds_j = \Gamma_i \oint_{C_i} \mathbf{A}_{\omega,j}(s_i) \cdot \mathbf{t}_i(s_i) ds_i.$$

An immediate corollary that follows, using (4.17) and (4.3), is that

$$\oint_{C_j} \mathbf{A}_{I,i}^u(s_j) \cdot \mathbf{t}_j(s_j) ds_j = \oint_{C_i} \mathbf{A}_{I,j}^u(s_i) \cdot \mathbf{t}_i(s_i) ds_i.$$

Using this H can now be written as

$$2H(C_1,\cdots,C_N) := 2\underbrace{\sum_{\substack{i,j\\ i>j}}\Gamma_j\oint_{C_j}\mathbf{A}_{\omega,i}(s_j)\cdot\mathbf{t}_j(s_j)ds_j}_{H_a}+\underbrace{\sum\Gamma_j^2\oint_{C_j}\mathbf{A}^u_{I,j}(s_j)\cdot\mathbf{t}_j(s_j)ds_j}_{H_b}+2\sum H^j_{\mathrm{SI,reg}}. \tag{4.18}$$

Each variation $\delta C\in T_pP_{\mathrm{f}}^N$ can be uniquely represented by its N projected components, $(\delta C_1,\cdots,\delta C_N)$, with δC_j representing a curve-normal variation of the jth ring. Similarly,

$$X_H\equiv(X_{H_1},\cdots,X_{H_N}),\quad \frac{\delta F}{\delta C}\equiv\left(\frac{\delta F}{\delta C_1},\cdots,\frac{\delta F}{\delta C_N}\right),$$

where $\delta F/\delta C_j$ is paired with δC_j as per (2.29). Referring to (1.41), the Hamiltonian vector field X_H on P_{f}^N, equipped with the bracket (2.32), of any function $H:P_{\mathrm{f}}^N\to\mathbb{R}$ is then obtained as:

$$\dot F=\left\langle X_H,\frac{\delta F}{\delta C_j}\right\rangle_{P_{\mathrm{f}}^N}:=\sum\left\langle X_{H_j},\frac{\delta F}{\delta C_j}\right\rangle_{P_{\mathrm{f}}}=\sum\oint_{C_j}\left\langle X_{H_j},\left(\frac{\delta F}{\delta C_j}\right)^{\sharp}\right\rangle_{\mathbb{R}^3}ds$$
$$=\frac{1}{\Gamma_j}\oint_{C_j}\mathbf{t}_j\cdot\left(\left(\frac{\delta F}{\delta C_j}\right)^{\sharp}\times\left(\frac{\delta H}{\delta C_j}\right)^{\sharp}\right)ds.$$

The last equation gives

$$X_{H_j}=\frac{1}{\Gamma_j}\left(\frac{\delta H}{\delta C_j}\right)^{\sharp}\times\mathbf{t}_j. \tag{4.19}$$

Recall from (4.3) that $\mathbf{A}_i$ is the vector potential field of the ith ring. Applying (2.29) and (2.12), obtain

$$\left\langle\frac{\delta H_a}{\delta C_j},\delta C_j\right\rangle$$
$$=\lim_{\varepsilon\to 0}\frac{1}{\varepsilon}\left[\Gamma_j\oint_{C_j}\left(\left(\mathbf{A}_{\omega,i}+\varepsilon\delta\mathbf{A}_{\omega,i}\right)(s_j)\cdot\mathbf{t}_j(s_j)+\mathbf{A}_{\omega,i}(s_j)\cdot\left(\mathbf{t}_j+\varepsilon\delta\mathbf{t}_j\right)(s_j)\right)ds_j\right.$$
$$\left.-\Gamma_j\oint_{C_j}\mathbf{A}_{\omega,i}(s_j)\cdot\mathbf{t}_j(s_j)ds_j\right],$$
$$=\Gamma_j\oint_{C_j}\left(\left(\mathbf{D_r}\mathbf{A}_{\omega,i}\cdot\delta\mathbf{r}_j\right)(s_j)\cdot\mathbf{t}_j(s_j)+\mathbf{A}_{\omega,i}(s_j)\cdot\delta\mathbf{t}_j(s_j)\right)ds_j,$$

where $\mathbf{D_r}\mathbf{A}_{\omega,i}$ is the 3×3 Jacobian matrix of the map $\mathbf{A}_{\omega,i}:D\to\mathbb{R}^3$. Since s_j is the arc-length parameter, $\mathbf{t}_j=d\mathbf{r}_j/ds_j$. Making the usual assumption of variations commuting with derivatives, $\delta \mathbf{t}_j$ can be replaced by $d\delta\mathbf{r}_j/ds_j$. Integration by parts then allows one to write,

$$\begin{aligned}\left\langle \frac{\delta H_a}{\delta C_j}, \delta C_j \right\rangle &= \Gamma_j \oint_{C_j} \left((\mathbf{D_r}\mathbf{A}_{\omega,i} \cdot \delta\mathbf{r}_j)(s_j) \cdot \mathbf{t}_j(s_j) - \frac{d}{ds_j}(\mathbf{A}_{\omega,i}(s_j)) \cdot \delta\mathbf{r}_j(s_j) \right) ds_j, \\ &= \Gamma_j \oint_{C_j} ((\mathbf{D_r}\mathbf{A}_{\omega,i} \cdot \delta\mathbf{r}_j)(s_j) \cdot \mathbf{t}_j(s_j) - (\mathbf{D_r}\mathbf{A}_{\omega,i} \cdot \mathbf{t}_j)(s_j) \cdot \delta\mathbf{r}_j(s_j)) ds_j, \\ &= \Gamma_j \oint_{C_j} \left(\left((\mathbf{D_r}\mathbf{A}_{\omega,i})^T - \mathbf{D_r}\mathbf{A}_{\omega,i} \right) \cdot \mathbf{t}_j \right)(s_j) \cdot \delta\mathbf{r}_j(s_j) ds_j, \\ &= \Gamma_j \oint_{C_j} ((-\nabla \times \mathbf{A}_{\omega,i}) \times \mathbf{t}_j)(s_j) \cdot \delta\mathbf{r}_j(s_j) ds_j.\end{aligned} \tag{4.20}$$

In the last line, use is made of the isomorphism between 3×3 skew-symmetric matrices and vectors in $\mathbb{R}^3$. And so,

$$\left(\frac{\delta H_a}{\delta C_j} \right)^\sharp = \Gamma_j (-\nabla \times \mathbf{A}_{\omega,i}) \times \mathbf{t}_j.$$

To compute $\delta H_b / \delta C_j$, one can invoke the following argument instead of going through a computation. Referring to the right side of Eq. (4.20), the integrand is the differential change obtained by varying the position of ring C_j while keeping ring C_i fixed. It is easy to see that the differential change due to varying the position of ring C_i while keeping ring C_j fixed is obtained by simply replacing $\delta\mathbf{r}_j$ by $\delta\mathbf{r}_i$. The differential change due to varying C_i and C_j *simultaneously* is then equal to the sum of these two changes. Since H_b is obtained by setting $i = j$ it follows that

$$\left(\frac{\delta H_b}{\delta C_j} \right)^\sharp = 2\Gamma_j^2 (-\nabla \times \mathbf{A}^u_{I,j}) \times \mathbf{t}_j.$$

Gathering all terms, the functional derivative of H with respect to C_j takes the form

$$\left(\frac{\delta H}{\delta C_j} \right)^\sharp = \sum_{i \neq j} \Gamma_j (-\nabla \times \mathbf{A}_{\omega,i}) \times \mathbf{t}_j + \Gamma_j^2 (-\nabla \times \mathbf{A}^u_{I,j}) \times \mathbf{t}_j + \left(\frac{\delta H^j_{\mathrm{SI,reg}}}{\delta C_j} \right)^\sharp. \tag{4.21}$$

From (4.19) it follows that

$$\begin{aligned} X_{H_j} &= \left(\sum_{i \neq j} (-\nabla \times \mathbf{A}_{\omega,i}) \times \mathbf{t}_j + \Gamma_j (-\nabla \times \mathbf{A}^u_{I,j}) \times \mathbf{t}_j + \frac{1}{\Gamma_j} \left(\frac{\delta H^j_{\mathrm{SI,reg}}}{\delta C_j} \right)^\sharp \right) \times \mathbf{t}_j, \\ &= \sum_{i \neq j} (\mathbf{v}_i)_{n_j} + \Gamma_j (\mathbf{v}_{I,j})_{n_j} + \frac{1}{\Gamma_j} \left(\frac{\delta H^j_{\mathrm{SI,reg}}}{\delta C_j} \right)^\sharp \times \mathbf{t}_j, \end{aligned}$$

where in the first two terms the operation $\mathbf{t}_j \times ((\cdots) \times \mathbf{t}_j)$ is identified with the projection onto the normal plane. From (4.4), after making the appropriate changes in the indices, one finally obtains that (for each $j = 1, \cdots, N$) X_{H_j} is equal to the right hand side of (4.15), thus proving the Proposition.

For future reference, it is useful to write the equations of this system in Hamiltonian form as

$$X_H \equiv \left(\frac{1}{\Gamma_1} \left(\frac{\delta H}{\delta C_1} \right)^\sharp \times \mathbf{t}_1, \cdots, \frac{1}{\Gamma_N} \left(\frac{\delta H}{\delta C_N} \right)^\sharp \times \mathbf{t}_N \right) = \left(\frac{\partial C_1}{\partial t}, \cdots, \frac{\partial C_N}{\partial t} \right), \quad (4.22)$$

with H given by (4.16).

Euler–Lagrange Formulation for the Stationary Body Problem

Recall that in Chap. 2, the degenerate Lagrangian construction of Chapman for the N-point-vortex problem was applied to the N vortex rings model. This construction can also be applied here. Going over the derivation, starting from Eq. (2.37) to Eq. (2.40), it is easily seen that in these equations the H function can be replaced by any smooth function of $(C_1, \cdots, C_N)$ leading to the same final form of Eq. (2.40). Therefore, in this problem the Lagrangian $\mathscr{L}_\mathrm{f} : T_p P_\mathrm{f}^N \to \mathbb{R}$ is given by

$$\mathscr{L}_\mathrm{f}\left(C_1, \cdots, C_N, \frac{\partial C_1}{\partial t}, \cdots, \frac{\partial C_N}{\partial t} \right)$$
$$:= -\frac{1}{3} \sum \Gamma_j \oint_{C_j} \left\langle \mathbf{t}_j(s_j), C_j(s_j) \times \frac{\partial C_j}{\partial t}(s_j) \right\rangle_{\mathbb{R}^3} ds_j - H(C_1, \cdots, C_N), \quad (4.23)$$

with H given by (4.16), and the Euler–Lagrange equations obtained then are exactly the same as (4.22).

4.2 N Vortex Rings Dynamically Coupled with a Single Rigid Body of Arbitrary Shape

The model of the previous subsection is now extended to the case where the body is free to move under the action of the fluid stresses. This model thus also becomes the $\mathbb{R}^3$-extension of the model of Sect. 3.2. As in the planar case, one can also view this as an extension of Kirchhoff's original model in $\mathbb{R}^3$ with the addition of vortex rings. A schematic sketch is shown in Fig. 4.1. Due to the inherent SE(3) symmetries, the system dynamics is again independent of the absolute location and orientation of the body or the rings in $\mathbb{R}^3$, and depends only on the relative positioning of all these objects. The choice of a body-fixed frame is again appropriate. Full details of the equations of motion and the Hamiltonian formalism may be found in [157], only the main features are recalled here.

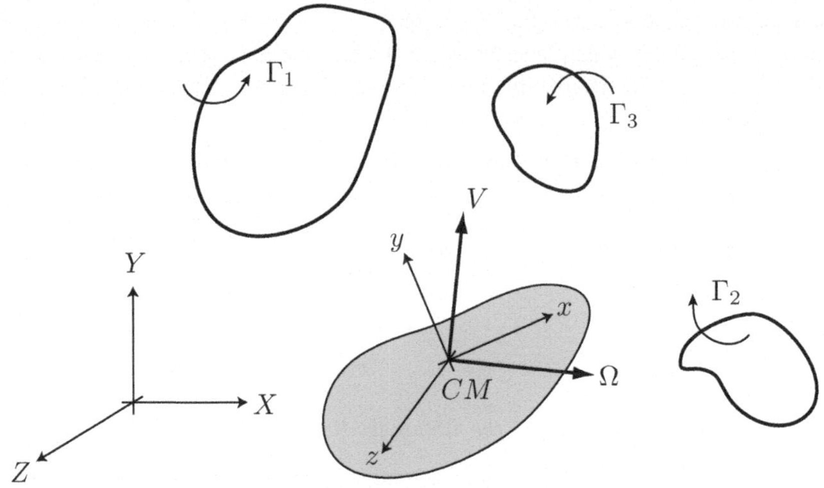

Fig. 4.1 A neutrally buoyant rigid body of arbitrary smooth shape dynamically interacting with vortex rings of arbitrary strengths and signs external to it (from [157])

The motion of the body introduces a Kirchhoff velocity field $\mathbf{u}_B = \nabla \Phi_B$, where Φ_B is given by (1.4). The Hodge–Helmholtz decomposition (3.12) continues to apply along with the boundary conditions in (1.3), (3.1), and (3.13). In the latter equation note that now $\mathbf{v}_o = \sum \mathbf{v}_j$. The system momenta variables referred to the body-fixed frame, $\mathbf{L}$ and $\mathbf{A}$, are given by

$$\begin{pmatrix} \mathbf{L} \\ \mathbf{A} \end{pmatrix} = (\mathbf{M_b} + \mathbf{M_a}) \cdot \begin{pmatrix} \mathbf{V} \\ \Omega \end{pmatrix} + \begin{pmatrix} \mathbf{P} \\ \Pi \end{pmatrix}. \tag{4.24}$$

with $\mathbf{M_b}, \mathbf{M_a}, \mathbf{V}$, and Ω representing the same physical quantities as before, but with $\mathbf{M_b} + \mathbf{M_a}$ now a 6×6 symmetric matrix and with $\mathbf{V}$ and Ω vectors in $\mathbb{R}^3$. The terms representing the vortical momenta of the rings are now given by

$$\mathbf{P} = \frac{1}{2} \sum \Gamma_j \left(\oint_{C_j} (\mathbf{l}_j(s_j) \times \mathbf{t}_j(s_j)) ds_j \right) + \frac{1}{2} \int_{\partial B} \mathbf{l} \times (\mathbf{n} \times \mathbf{v}_\omega)\, \nu, \tag{4.25}$$

$$\Pi = -\frac{1}{2} \sum \Gamma_j \oint_{C_j} l_j^2(s_j) \mathbf{t}_j(s_j) ds_j - \frac{1}{2} \int_{\partial B} l^2 (\mathbf{n} \times \mathbf{v}_\omega)\, \nu, \tag{4.26}$$

where $\mathbf{l}$ denotes the position vector of points in the body-fixed frame, $l = \|\mathbf{l}\|$, $\mathbf{l}_j$ is $\mathbf{l}$ for points on the jth ring with $l_j = \|\mathbf{l}_j\|$. Referring to Eq. (2.12), the corresponding notational equivalence in this body-fixed frame setting is $\mathbf{l}_j(s_j) \equiv C_j(s_j)$. It is easy to see from (3.1), (3.13), and (4.4) that for a given body shape, the body integrals in $\mathbf{P}$ and Π depend only on the relative distribution of the rings around the body and their strengths.

The equations of motion, referred to a body-fixed frame, of a rigid body of arbitrary smooth shape, $B \subset \mathbb{R}^3$, dynamically interacting with N vortex rings external to it are

$$\left(\frac{d}{dt}+\Omega\times\right)\mathbf{L}=0, \tag{4.27}$$

$$\left(\frac{d}{dt}+\Omega\times\right)\mathbf{A}+\mathbf{V}\times\mathbf{L}=0, \tag{4.28}$$

$$\frac{\partial C_j}{\partial t}+(\mathbf{V}+\Omega\times\mathbf{l}_j)_{|n_j}=\left(\sum_{i\neq j}\mathbf{v}_{\omega,i}+\mathbf{v}_{I,j}+\nabla\Phi_B\right)_{|n_j}+\left(\mathbf{v}_{\mathrm{SI},j}\right)_{|n_j},\quad j=1,\cdots,N. \tag{4.29}$$

Proposition 4.2. ([157]) *The system of Eqs. (4.27), (4.28), and (4.29) with Hamiltonian function*

$$\begin{aligned}
&H(\mathbf{L},\mathbf{A},C_1,\cdots,C_N)\\
&=\frac{1}{2}\sum_{\substack{i,j\\ i\neq j}}\Gamma_j\oint_{C_j}\mathbf{A}_i(s_j)\cdot\mathbf{t}_j(s_j)ds_j+\sum_{i,j}\Gamma_i\Gamma_j\oint_{C_j}\mathbf{A}^u_{I,i}(s_j)\cdot\mathbf{t}_j(s_j)ds_j\\
&\qquad+\frac{1}{2}\sum H^j_{\mathrm{SI,reg}}+\frac{1}{2}\left\langle(\mathbf{L},\mathbf{A})-(\mathbf{P},\Pi),(\mathbf{M_b}+\mathbf{M_a})^{-1}((\mathbf{L},\mathbf{A})-(\mathbf{P},\Pi))\right\rangle,
\end{aligned} \tag{4.30}$$

is a Poisson vector field on the space $P=\mathfrak{se}(3)^*\times U_{\mathrm{f}}^N$, where

$$U_{\mathrm{f}}^N:=\underbrace{(P_{\mathrm{f}}\times\cdots\times P_{\mathrm{f}})}_{N\text{ times}}\setminus\left(\triangle\cup\triangle_b^N\right)\equiv P_{\mathrm{f}}^N\setminus\triangle_b^N,$$

$\triangle_b$ denoting the set of points corresponding to intersections of each ring with the body. The Poisson bracket is defined as follows. For $F,G\in C^\infty(P)$, define, following (1.47),

$$\{F,G\}_P=\{F_1,G_1\}_{\mathfrak{se}(3)^*}+\{F_2,G_2\}_{\mathrm{f}}, \tag{4.31}$$

where the second bracket is given by (2.32) on U_{f}^N. Therefore, as per (1.41), (1.36), and (2.18), if $p(t)=(\mu(t),C_1(t),\cdots,C_N(t))\in P$ is an integral curve of the system, where $\mu(t)=(\mathbf{L}(t),\mathbf{A}(t))$, then it satisfies

$$\begin{aligned}
\frac{dF}{dt}:=\left\langle\nabla_pF,\frac{\partial p}{dt}\right\rangle&\equiv\left\langle\frac{\partial F}{\partial\mu},\frac{d\mu}{dt}\right\rangle_{\mathfrak{se}(3)}+\sum\oint_{C_j}\left\langle\left(\frac{\delta F}{\delta C_j}\right)^\sharp,\frac{\partial C_j}{\delta t}\right\rangle_{\mathbb{R}^3}ds_j\\
&=\left\langle\frac{\partial F}{\partial\mu},\mathrm{ad}^*_{\partial H/\partial\mu}\mu\right\rangle_{\mathfrak{se}(3)}\\
&\qquad+\sum\frac{1}{\Gamma_j}\oint_{C_j}\left\langle\mathbf{t}_j,\left(\frac{\delta F}{\delta C_j}\right)^\sharp\times\left(\frac{\delta H}{\delta C_j}\right)^\sharp\right\rangle_{\mathbb{R}^3}ds_j,
\end{aligned}$$

leading to Eqs. (4.27), (4.28), and (4.29).

Remarks

1. The rings bracket (2.32) is SE(3) invariant and shows up in the same form in this model.
2. Unlike the point vortex model, this model is an infinite-dimensional model due to the presence of the rings.
3. Consistent with the bracket structure, Eqs. (4.29) can be written in the form

$$\frac{\partial C_j}{\partial t} = \frac{1}{\Gamma_j}\left(\frac{\delta H}{\delta C_j}\right)^{\sharp} \times \mathbf{t}_j, \quad j = 1, \cdots, N \tag{4.32}$$

with H given by (4.30). This is shown toward the end of Sect. 4.2.2.
4. Similar to the point vortex model, the following features are noted (in this item and the next two). The variables $(\mathbf{L}, \mathbf{A})$ and the variables C_j can be varied independently of each other. This is easily seen from (4.58). Any variation $\delta C_j, j = 1, \cdots, N$, with $\delta(\mathbf{L}, \mathbf{A}) = 0$, can be achieved by varying $(\mathbf{V}, \Omega)$ so that $\delta(\mathbf{V}, \Omega) = -(\mathbf{M_b} + \mathbf{M_a})^{-1} \cdot \delta(\mathbf{P}, \Pi)$. Similarly, any variation $\delta(\mathbf{L}, \mathbf{A})$, with $\delta C_j = 0, j = 1, \cdots, N$, and hence $\delta(\mathbf{P}, \Pi) = 0$, can be achieved by varying $(\mathbf{V}, \Omega)$ such that $\delta(\mathbf{V}, \Omega) = -(\mathbf{M_b} + \mathbf{M_a})^{-1} \cdot \delta(\mathbf{L}, \mathbf{A})$.
5. Simplifications occur when the rigid body has a simple geometry. In particular, for the case of a sphere, which will be discussed more later, the evolution equation for $\mathbf{A}$ essentially decouples from the other equations.
6. For an arbitrary (but smooth) body shape, the image velocity fields and the Kirchhoff unit potentials may not have closed form expressions. In such a case, Eqs. (4.27)–(4.29) have to be appended by Eq. (3.1) for the image velocity field, at each time instant t, and by Eqs. (1.5) and (1.6) for the unit potentials. And the latter two sets of equations would have to be solved (numerically) only once.
7. As stated in the introduction to this chapter, owing to the fact that the total vorticity is zero, this Hamiltonian structure holds regardless of whether $\Gamma := \sum \Gamma_j$ is zero or not.

Conserved Quantities

From Remark 4, one has $\partial \mathbf{L}/\partial C_j = 0$ and it follows that $\mathscr{C}_1$ and $\mathscr{C}_2$ given by (1.42) are also Casimirs of the bracket (4.31), with $(\mathbf{L}, \mathbf{A})$ now given by (4.58). The fact that these are conserved quantities can be easily verified from the form of Eqs. (4.27) and (4.28). These conserved quantities may be expressed in terms of the body velocities as:

$$\Big\langle \mathbf{M}_{11} \cdot \mathbf{V}(t) + \mathbf{M}_{12} \cdot \Omega(t), \mathbf{M}_{11} \cdot \mathbf{V}(t) + \mathbf{M}_{12} \cdot \Omega(t) \Big\rangle_{\mathbb{R}^3}$$
$$+ 2\Big\langle \mathbf{M}_{11} \cdot \mathbf{V}(t) + \mathbf{M}_{12} \cdot \Omega(t), \mathbf{P}(t) \Big\rangle_{\mathbb{R}^3} + \Big\langle \mathbf{P}(t), \mathbf{P}(t) \Big\rangle_{\mathbb{R}^3} = \text{constant},$$

$$\left\langle \mathbf{M}_{11} \cdot \mathbf{V}(t) + \mathbf{M}_{12} \cdot \Omega(t), \mathbf{M}_{12}^T \cdot \mathbf{V}(t) + \mathbf{M}_{22} \cdot \Omega(t) \right\rangle_{\mathbb{R}^3}$$
$$+ \left\langle \mathbf{M}_{11} \cdot \mathbf{V}(t) + \mathbf{M}_{12} \cdot \Omega(t), \Pi(t) \right\rangle_{\mathbb{R}^3} + \left\langle \mathbf{M}_{12}^T \cdot \mathbf{V}(t) + \mathbf{M}_{22} \cdot \Omega(t), \mathbf{P}(t) \right\rangle_{\mathbb{R}^3}$$
$$+ \left\langle \mathbf{P}(t), \Pi(t) \right\rangle_{\mathbb{R}^3} = \text{constant},$$

where the various matrix blocks are as defined in (1.14).

4.2.1 The Euler–Lagrange Equations in a Spatially-Fixed Frame

The Euler–Lagrange formulation for the dynamically coupled problem referred to a spatially-fixed frame is now presented. As in the planar problem, the Lagrangian is the sum of the Lagrangian in Kirchhoff's problem, (1.22), and the Lagrangian in the rings and stationary rigid body problem, (4.23), suitably modified:

$$\begin{aligned} &\mathscr{L}_{\mathrm{f}}\left(\mathbf{R}, \bar{\mathbf{r}}_c, \dot{\mathbf{R}}, \bar{\mathbf{V}}, \bar{C}_1, \cdots, \bar{C}_N, \frac{\partial \bar{C}_1}{\partial t}, \cdots, \frac{\partial \bar{C}_N}{\partial t}\right) \\ &:= \frac{1}{2}\langle \mathbf{V}, \mathbf{L}\rangle_{\mathbb{R}^3} + \frac{1}{2}\langle \Omega, \mathbf{A}\rangle_{\mathbb{R}^3} - \frac{1}{3} V_{\mathrm{f}}\left(\bar{C}_1, \cdots, \bar{C}_N, \frac{\partial \bar{C}_1}{\partial t}, \cdots, \frac{\partial \bar{C}_N}{\partial t}\right) \\ &\qquad - \bar{H}\left(\bar{C}_1, \cdots, \bar{C}_N, \mathbf{R}, \bar{\mathbf{r}}_c, \bar{\Omega}, \bar{\mathbf{V}}\right), \end{aligned} \tag{4.33}$$

where $\mathbf{L}$ and $\mathbf{A}$ are defined as in Kirchhoff's problem (1.12) and

$$V_{\mathrm{f}}\left(\bar{C}_1, \cdots, \bar{C}_N, \frac{\partial \bar{C}_1}{\partial t}, \cdots, \frac{\partial \bar{C}_N}{\partial t}\right) = \sum \Gamma_j \oint_{C_j} \left\langle \bar{\mathbf{t}}_j, \bar{C}_j \times \frac{\partial \bar{C}_j}{\partial t} \right\rangle_{\mathbb{R}^3} ds_j.$$

Referring to (2.12) and the lines below (4.26), it is noted that $\bar{C}_j$ and C_j are related by the equation:

$$\bar{\mathbf{r}}_j(s_j) = \mathbf{R} \cdot \mathbf{l}_j(s_j) + \bar{\mathbf{r}}_c$$

or, equivalently,

$$\bar{C}_j(s_j) = \mathbf{R} \cdot C_j(s_j) + \bar{\mathbf{r}}_c. \tag{4.34}$$

The function $\bar{H}$ is defined as follows. Referring to (4.18), let

$$\begin{aligned} H_G(C_1, \cdots, C_N) &:= H_a(C_1, \cdots, C_N) + \frac{H_b}{2}(C_1, \cdots, C_N) + \sum H_{\mathrm{SI,reg}}^j, \\ \bar{H}_G\left(\bar{C}_1, \cdots, \bar{C}_N, \mathbf{R}, \bar{\mathbf{r}}_c\right) &:= H_G(C_1, \cdots, C_N) \\ &\equiv H_G\left(\mathbf{R}^{-1} \cdot (\bar{C}_1 - \bar{\mathbf{r}}_c), \cdots, \mathbf{R}^{-1} \cdot (\bar{C}_N - \bar{\mathbf{r}}_c)\right). \end{aligned}$$

Now consider Eq. (4.46) from the next chapter, which introduces the vector potential field $\mathbf{A}_B(\mathbf{l},\mathbf{V},\Omega)$ which is the analog of the Kirchhoff flow streamfunction ψ_B. From this, define

$$\bar{\mathbf{A}}_B(\bar{\mathbf{r}},\bar{\mathbf{V}},\bar{\Omega}) := \mathbf{A}_B(\mathbf{l},\mathbf{V},\Omega) \equiv \mathbf{A}_B(\mathbf{R}^{-1}\cdot(\bar{\mathbf{r}}-\bar{\mathbf{r}}_c),\mathbf{R}^{-1}\cdot\bar{\mathbf{V}},\mathbf{R}^{-1}\cdot\bar{\Omega})$$

and

$$\begin{aligned} \bar{H}\left(\bar{C}_1,\cdots,\bar{C}_N,\mathbf{R},\bar{\mathbf{r}}_c,\bar{\Omega},\bar{\mathbf{V}}\right) &= K\left(C_1,\cdots,C_N,\Omega,\mathbf{V}\right), \\ &:= H_G\left(C_1,\cdots,C_N\right)+\sum\Gamma_j\oint_{C_j}\mathbf{A}_B(s_j)\cdot\mathbf{t}_j(s_j)ds_j, \\ &= \bar{H}_G\left(\bar{C}_1,\cdots,\bar{C}_N,\mathbf{R},\bar{\mathbf{r}}_c\right)+\sum\Gamma_j\oint_{\bar{C}_j}\bar{\mathbf{A}}_B(s_j)\cdot\bar{\mathbf{t}}_j(s_j)ds_j, \end{aligned} \tag{4.35}$$

where $\bar{H}^j_{\mathrm{SI,reg}}(\bar{C}_j) := H^j_{\mathrm{SI,reg}}(C_j)$. It is easily seen that

$$\frac{\delta\bar{H}}{\delta\bar{C}_j} = \mathbf{R}\cdot\left(\frac{\delta K}{\delta C_j}\right)^\sharp, \quad j=1,\cdots,N,$$

and that, moreover,

$$\left(\frac{\delta H_G}{\delta C_j}\right)^\sharp = \sum_{i\neq j}\Gamma_j(-\nabla\times\mathbf{A}_{\omega,i})\times\mathbf{t}_j+\Gamma_j^2\left(-\nabla\times\mathbf{A}^u_{I,j}\right)\times\mathbf{t}_j+\left(\frac{\delta H^j_{\mathrm{SI,reg}}}{\delta C_j}\right)^\sharp, \tag{4.36}$$

i.e. it is exactly the same as the right hand side of (4.21). Analogous to (4.20) one can also see that

$$\frac{\delta}{\delta C_j}\left(\sum\Gamma_j\oint_{C_j}\mathbf{A}_B(s_j)\cdot\mathbf{t}_j(s_j)ds_j\right)^\sharp = -\Gamma_j\nabla\Phi_B\times\mathbf{t}_j. \tag{4.37}$$

With these preliminaries established, the following proposition is now proved:

Proposition 4.3. *The Euler–Lagrange equations for the Lagrangian defined in (4.33) are equivalent to the system of Eqs. (4.27)–(4.29).*

Proof. As in Sects. 1.4 and 3.2.1, the least action principle is now implemented

$$\delta\int_{t_1}^{t_2}\mathscr{L}\left(\mathbf{R},\bar{\mathbf{r}}_c,\dot{\mathbf{R}},\bar{\mathbf{V}},\bar{C}_1,\cdots,\bar{C}_N,\frac{\partial\bar{C}_1}{\partial t},\cdots,\frac{\partial\bar{C}_N}{\partial t}\right)dt=0, \tag{4.38}$$

for variations taken over curves $\delta q(t) \equiv (\delta\mathbf{R}(t), \delta\bar{\mathbf{r}}_c(t), \delta\bar{C}_1(t), \cdots, \delta\bar{C}_N(t)) \in T_qQ$ satisfying the fixed end-points condition: $\delta\mathbf{R}(t_1) = \delta\mathbf{R}(t_2) = 0, \delta\bar{\mathbf{r}}_c(t_1) = \delta\bar{\mathbf{r}}_c(t_2) = 0$ and $\delta\bar{C}_j(t_1) = \delta\bar{C}_j(t_2) = 0\ (j = 1, \cdots, N)$.

Importing terms from (1.28) and from (2.39), after substituting terms with overhead bars in the latter, obtain

$$
\begin{aligned}
&\delta \int_{t_1}^{t_2} \mathscr{L}\left(\mathbf{R}, \bar{\mathbf{r}}_c, \dot{\mathbf{R}}, \bar{\mathbf{V}}, \bar{C}_1, \cdots, \bar{C}_N, \frac{\partial \bar{C}_1}{\partial t}, \cdots, \frac{\partial \bar{C}_N}{\partial t}\right) dt \\
&= \delta \int_{t_1}^{t_2} \left(\frac{1}{2}\langle \mathbf{V}, \mathbf{L}\rangle_{\mathbb{R}^3} + \frac{1}{2}\langle \Omega, \mathbf{A}\rangle_{\mathbb{R}^3} - \frac{1}{3}V_{\mathrm{f}}\left(\bar{C}_1, \cdots, \bar{C}_N, \frac{\partial \bar{C}_1}{\partial t}, \cdots, \frac{\partial \bar{C}_N}{\partial t}\right)\right. \\
&\qquad\qquad \left. - \bar{H}\left(\bar{C}_1, \cdots, \bar{C}_N, \mathbf{R}, \bar{\mathbf{r}}_c, \bar{\Omega}, \bar{\mathbf{V}}\right)\right) dt \\
&= \int_{t_1}^{t_2} \left(\left\langle \delta\bar{\mathbf{r}}_c, -\frac{d(\mathbf{R}\cdot\mathbf{L})}{dt} - \frac{\partial \bar{H}}{\partial \bar{\mathbf{r}}_c}\right\rangle_{\mathbb{R}^3} - \frac{1}{2}\operatorname{trace}\left(\delta\mathbf{R}^T \cdot \mathbf{R}\cdot \widehat{\bar{\mathbf{V}}\times \mathbf{L}}\right)\right. \\
&-\frac{1}{2}\operatorname{trace}\left(\delta\mathbf{R}^T \cdot \frac{d(\mathbf{R}\cdot\hat{\mathbf{A}})}{dt} - \delta\mathbf{R}^T\cdot\mathbf{R}\cdot\hat{\Omega}\cdot\hat{\mathbf{A}} + \delta\mathbf{R}^T\cdot\mathbf{R}\cdot\widehat{\bar{\Omega}\times\mathbf{A}}\right) \\
&\qquad - \left\langle \frac{\partial \bar{H}}{\partial \mathbf{R}}, \delta\mathbf{R}\right\rangle_{\mathrm{SO(3)}} - \left\langle \delta\bar{\mathbf{V}}, \frac{\partial \bar{H}}{\partial \bar{\mathbf{V}}}\right\rangle_{\mathbb{R}^3} - \left\langle \delta\bar{\Omega}, \frac{\partial \bar{H}}{\partial \bar{\Omega}}\right\rangle_{\mathbb{R}^3} \\
&\qquad\qquad \left. + \sum \oint_{\bar{C}_j} \left\langle \delta\bar{C}_j, \Gamma_j \bar{\mathbf{t}}_j \times \frac{\partial \bar{C}_j}{\partial t} - \left(\frac{\delta \bar{H}}{\delta \bar{C}_j}\right)^{\sharp}\right\rangle_{\mathbb{R}^3} (s_j)\, ds_j\right) dt,
\end{aligned}
$$

where the fact that $\bar{H}$ is independent of $\partial C_j/\partial t$ has already been used.

The variational derivatives of $\bar{H}$ with respect to $\bar{\mathbf{r}}_c, \bar{C}_1, \cdots, \bar{C}_N, \bar{\mathbf{V}}, \bar{\Omega}$ and $\mathbf{R}$ are now computed. As per (4.34), a variation $\bar{\mathbf{r}}_c$, keeping $\bar{C}_j$ fixed, induces a variation δC_j in each ring. From (4.35), these induced variations have to satisfy

$$
\begin{aligned}
\left\langle \delta\bar{\mathbf{r}}_c, \frac{\partial \bar{H}}{\partial \bar{\mathbf{r}}_c}\right\rangle_{\mathbb{R}^3} &= \sum \oint_{C_j} \left\langle \delta C_j, \left(\frac{\delta K}{\delta C_j}\right)^{\sharp}\right\rangle_{\mathbb{R}^3} (s_j)\, ds_j \\
&= -\sum \oint_{C_j} \left\langle \mathbf{R}^{-1}\cdot\delta\bar{\mathbf{r}}_c, \left(\frac{\delta K}{\delta C_j}\right)^{\sharp}\right\rangle_{\mathbb{R}^3} (s_j)\, ds_j \\
\Rightarrow \frac{\partial \bar{H}}{\partial \bar{\mathbf{r}}_c} &= -\mathbf{R}\cdot\sum \oint_{C_j} \left(\frac{\delta K}{\delta C_j}\right)^{\sharp} (s_j)\, ds_j.
\end{aligned}
$$

From $\bar{\mathbf{V}} = \mathbf{R}\cdot\mathbf{V}$ and $\bar{\Omega} = \mathbf{R}\cdot\Omega$, it follows that

$$\frac{\partial \bar{H}}{\partial \bar{\mathbf{V}}} = \mathbf{R}\cdot\frac{\partial K}{\partial \mathbf{V}} = \mathbf{R}\cdot\sum\Gamma_j\oint_{C_j}[A_T](s_j)\cdot\mathbf{t}_j(s_j)\,ds_j = \mathbf{R}\cdot -\frac{1}{2}\int_{\partial B}\mathbf{l}\times(\mathbf{n}\times\mathbf{v}_\omega)\,\nu, \tag{4.39}$$

$$\frac{\partial \bar{H}}{\partial \bar{\Omega}} = \mathbf{R}\cdot\frac{\partial K}{\partial \Omega} = \mathbf{R}\cdot\sum\Gamma_j\oint_{C_j}[A_R](s_j)\cdot\mathbf{t}_j(s_j)\,ds_j = \mathbf{R}\cdot\frac{1}{2}\int_{\partial B}l^2(\mathbf{n}\times\mathbf{v}_\omega)\,\nu, \tag{4.40}$$

where $[A_T]$ and $[A_R]$ are 3×3 matrices comprised of the first three columns and the last three columns of matrix $[A]$ in Eq. (4.46), respectively. The last equation in each line is obtained from using the reciprocity relations (4.51).

Variations in $\mathbf{R}$, keeping $\bar{C}_1,\cdots,\bar{C}_N,\bar{\mathbf{r}}_c,\bar{\Omega}$ and $\bar{\mathbf{V}}$ fixed, induce variations in $C_1,\cdots,C_N,\Omega$, and $\mathbf{V}$ leading to the following relations:

$$\begin{aligned}
&\left\langle\frac{\partial \bar{H}}{\partial \mathbf{R}},\delta\mathbf{R}\right\rangle_{\mathrm{SO}(3)} \\
&=\sum\oint_{C_j}\left\langle\delta C_j,\left(\frac{\delta K}{\delta C_j}\right)^\sharp\right\rangle_{\mathbb{R}^3}(s_j)ds_j+\left\langle\frac{\partial K}{\partial \mathbf{V}},\delta\mathbf{V}\right\rangle_{\mathbb{R}^3}+\left\langle\frac{\partial K}{\partial \Omega},\delta\Omega\right\rangle_{\mathbb{R}^3}, \\
&=\sum\oint_{C_j}\left\langle-\mathbf{R}^{-1}\cdot\delta\mathbf{R}\cdot C_j,\left(\frac{\delta K}{\delta C_j}\right)^\sharp\right\rangle_{\mathbb{R}^3}(s_j)ds_j+\left\langle\frac{\partial K}{\partial \mathbf{V}},-\mathbf{R}^{-1}\cdot\delta\mathbf{R}\cdot\mathbf{V}\right\rangle_{\mathbb{R}^3} \\
&\qquad\qquad+\frac{1}{2}\,\mathrm{trace}\left\langle\widehat{\frac{\partial K}{\partial \Omega}}^T\cdot-\mathbf{R}^{-1}\cdot\delta\mathbf{R}\cdot\hat{\Omega}\right\rangle, \\
&=-\frac{1}{2}\,\mathrm{trace}\left\langle\delta\mathbf{R}^T\cdot\mathbf{R}\cdot\sum\oint_{C_j}\widehat{C_j\times\left(\frac{\delta K}{\delta C_j}\right)^\sharp(s_j)ds_j}\right\rangle-\frac{1}{2}\,\mathrm{trace}\left\langle\delta\mathbf{R}^T\cdot\mathbf{R}\cdot\widehat{\mathbf{V}\times\frac{\partial K}{\partial \mathbf{V}}}\right\rangle \\
&+\frac{1}{2}\,\mathrm{trace}\left\langle\delta\mathbf{R}^T\cdot\mathbf{R}\cdot\left(\hat{\Omega}\cdot\widehat{\frac{\partial K}{\partial \Omega}}-\widehat{\Omega\times\frac{\partial K}{\partial \Omega}}\right)\right\rangle.
\end{aligned}$$

It should be noted that the relation $\delta C_j = -\mathbf{R}^{-1}\cdot\delta\mathbf{R}\cdot C_j$ used in the above equations, as also in the relation $\delta C_j = -\mathbf{R}^{-1}\cdot\delta\bar{\mathbf{r}}_c$ on the previous page, the right hand sides must account for the fact that only curve-normal variations δC_j are allowed. As will be seen later, the paired term $\delta K/\delta C_j$ obtained from the Euler–Lagrange equations is already a curve-normal component and so this filtering is naturally taken care of.

Finally the pairing with $\delta\bar{\Omega}$ is converted to a pairing with $\delta\dot{\mathbf{R}}$ using (1.21),

$$\begin{aligned}
\left\langle\delta\bar{\Omega},\frac{\partial \bar{H}}{\partial \bar{\Omega}}\right\rangle_{\mathbb{R}^3}=\left\langle\delta\Omega,\frac{\partial K}{\partial \Omega}\right\rangle_{\mathbb{R}^3}&=\frac{1}{2}\,\mathrm{trace}\left\langle\delta\hat{\Omega}^T\cdot\widehat{\frac{\partial K}{\partial \Omega}}\right\rangle, \\
&=\frac{1}{2}\,\mathrm{trace}\left\langle\delta\dot{\mathbf{R}}^T\cdot\mathbf{R}\cdot\widehat{\frac{\partial K}{\partial \Omega}}\right\rangle.
\end{aligned}$$

Gathering all these terms and using integration by parts and the fixed end-points conditions on pairings involving $\delta\bar{\mathbf{V}}$ and $\delta\dot{\mathbf{R}}$, the variation of the action now becomes:

$$
\begin{aligned}
&\delta \int_{t_1}^{t_2} \mathscr{L}\left(\mathbf{R}, \bar{\mathbf{r}}_c, \dot{\mathbf{R}}, \bar{\mathbf{V}}, \bar{C}_1, \cdots, \bar{C}_N, \frac{\partial \bar{C}_1}{\partial t}, \cdots, \frac{\partial \bar{C}_N}{\partial t}\right) dt \\
&= \int_{t_1}^{t_2} \left(\left\langle \delta \bar{\mathbf{r}}_c, -\frac{d(\mathbf{R}\cdot\mathbf{L})}{dt} + \mathbf{R}\cdot \sum \oint_{C_j} \left(\frac{\delta K}{\delta C_j}\right)^{\sharp} (s_j)\, ds_j + \frac{d}{dt}\left(\mathbf{R}\cdot\frac{\partial K}{\partial \mathbf{V}}\right)\right\rangle_{\mathbb{R}^3} \right.\\
&-\frac{1}{2}\operatorname{trace}\left(\delta \mathbf{R}^T\cdot\mathbf{R}\cdot\widehat{\mathbf{V}\times\mathbf{L}}\right) \\
&\qquad -\frac{1}{2}\operatorname{trace}\left(\delta\mathbf{R}^T\cdot\frac{d(\mathbf{R}\cdot\hat{\mathbf{A}})}{dt} - \delta\mathbf{R}^T\cdot\mathbf{R}\cdot\hat{\Omega}\cdot\hat{\mathbf{A}} + \delta\mathbf{R}^T\cdot\mathbf{R}\cdot\widehat{\Omega\times\mathbf{A}}\right) \\
&+\frac{1}{2}\operatorname{trace}\left\langle \delta\mathbf{R}^T\cdot\mathbf{R}\cdot\widehat{\sum\oint_{C_j} C_j\times\left(\frac{\delta K}{\delta C_j}\right)^{\sharp}(s_j)ds_j}\right\rangle + \frac{1}{2}\operatorname{trace}\left\langle\delta\mathbf{R}^T\cdot\mathbf{R}\cdot\widehat{\mathbf{V}\times\frac{\partial K}{\partial\mathbf{V}}}\right\rangle \\
&-\frac{1}{2}\operatorname{trace}\left\langle\delta\mathbf{R}^T\cdot\mathbf{R}\cdot\left(\hat{\Omega}\cdot\widehat{\frac{\partial K}{\partial\Omega}} - \widehat{\Omega\times\frac{\partial K}{\partial\Omega}}\right)\right\rangle + \frac{1}{2}\operatorname{trace}\left\langle\delta\mathbf{R}^T\cdot\frac{d}{dt}\left(\mathbf{R}\cdot\widehat{\frac{\partial K}{\partial\Omega}}\right)\right\rangle \\
&\qquad \left. +\sum\oint_{\bar{C}_j}\left\langle\delta\bar{C}_j, \Gamma_j\bar{\mathbf{t}}_j\times\frac{\partial\bar{C}_j}{\partial t} - \left(\frac{\delta\bar{H}}{\delta\bar{C}_j}\right)^{\sharp}\right\rangle_{\mathbb{R}^3}(s_j)\, ds_j\right) dt.
\end{aligned}
$$

Implementing the action principle and setting all terms paired with $\delta\bar{\mathbf{r}}_c$, $\delta\mathbf{R}$, and $\delta\bar{C}_1, \cdots, \delta\bar{C}_N$, respectively, to zero one obtains the following equations:

$$
\begin{aligned}
&\mathbf{R}\cdot\left[\left(\frac{d}{dt}+\Omega\times\right)\left(\mathbf{L}-\frac{\partial K}{\partial\mathbf{V}}\right) - \sum\oint_{C_j}\left(\frac{\delta K}{\delta C_j}\right)^{\sharp}(s_j)\, ds_j\right] = 0, \\
&\mathbf{R}\cdot\left[\frac{d}{dt}\left(\hat{\mathbf{A}} - \widehat{\frac{\partial K}{\partial\Omega}}\right) + \widehat{\Omega\times\left(\mathbf{A}-\frac{\partial K}{\partial\Omega}\right)} + \widehat{\mathbf{V}\times\left(\mathbf{L}-\frac{\partial K}{\partial\mathbf{V}}\right)}\right. \\
&\qquad\qquad \left. -\widehat{\sum\oint_{C_j} C_j\times\left(\frac{\delta K}{\delta C_j}\right)^{\sharp}(s_j)\, ds_j}\right] = 0, \\
&\mathbf{R}\cdot\left[\Gamma_j\mathbf{t}_j\times\left(\frac{\partial C_j}{\partial t} + \Omega\times C_j + \mathbf{V}\right) - \left(\frac{\delta K}{\delta C_j}\right)^{\sharp}\right] = 0, \quad j = 1, \cdots, N.
\end{aligned}
\tag{4.41}
$$

As mentioned previously, the equations in the last line already express the functional derivative in the curve-normal directions. Moreover, using (4.36) and (4.37) it follows that the last line of equations in (4.41) is identical to (4.29).

Substitute for $(\delta K/\delta C_j)^\sharp$ in the linear momentum equation in (4.41) and resolve the integral terms as follows. First, note that

$$\oint_{C_j} \frac{\partial}{\partial t}(\mathbf{t}_j \times C_j)\, ds_j = \oint_{C_j} \left(\frac{\partial \mathbf{t}_j}{\partial t} \times C_j\right) ds_j + \oint_{C_j} \left(\mathbf{t}_j \times \frac{\partial C_j}{\partial t}\right) ds_j,$$
$$= -\oint_{C_j} \left(\frac{\partial C_j}{\partial t} \times \mathbf{t}_j\right) ds_j + \oint_{C_j} \left(\mathbf{t}_j \times \frac{\partial C_j}{\partial t}\right) ds_j,$$
$$\Rightarrow \frac{1}{2}\oint_{C_j} \frac{\partial}{\partial t}(\mathbf{t}_j \times C_j)\, ds_j = \oint_{C_j} \left(\mathbf{t}_j \times \frac{\partial C_j}{\partial t}\right) ds_j, \tag{4.42}$$

after making use of arc-length parametrization, commutability of derivatives, and integration by parts. Next, note that

$$\begin{aligned}\oint_{C_j} \frac{\partial}{\partial s_j}(C_j \otimes C_j)\, ds_j = 0 &= \oint_{C_j} \frac{\partial C_j}{\partial s_j} \otimes C_j\, ds_j + \oint_{C_j} C_j \otimes \frac{\partial C_j}{\partial s_j}\, ds_j,\\ &= \oint_{C_j} \mathbf{t}_j \otimes C_j\, ds_j + \oint_{C_j} C_j \otimes \mathbf{t}_j\, ds_j,\end{aligned} \tag{4.43}$$

from which (and using standard vector identities) it follows that

$$\begin{aligned}\oint_{C_j} \mathbf{t}_j \times (\Omega \times C_j)\, ds_j &= -\Omega \times \left(\oint_{C_j} C_j \times \mathbf{t}_j\, ds_j\right) - \oint_{C_j} C_j \times (\mathbf{t}_j \times \Omega)\, ds_j,\\
&= -\frac{1}{2}\Omega \times \left(\oint_{C_j} C_j \times \mathbf{t}_j\, ds_j\right) - \frac{1}{2}\Omega \times \left(\oint_{C_j} C_j \times \mathbf{t}_j\, ds_j\right)\\
&\qquad + \Omega \oint_{C_j} C_j \cdot \mathbf{t}_j\, ds_j - \oint_{C_j} (C_j \cdot \Omega)\mathbf{t}_j\, ds_j\\
&= -\frac{1}{2}\Omega \times \left(\oint_{C_j} C_j \times \mathbf{t}_j\, ds_j\right) - \frac{1}{2}\oint_{C_j} (\Omega \cdot \mathbf{t}_j) C_j\, ds_j\\
&\qquad + \frac{\Omega}{2}\oint_{C_j} \frac{\partial(|C_j|^2)}{\partial s_j}\, ds_j - \frac{1}{2}\oint_{C_j} (C_j \cdot \Omega)\mathbf{t}_j\, ds_j,\\
&= -\frac{1}{2}\Omega \times \left(\oint_{C_j} C_j \times \mathbf{t}_j\, ds_j\right)\\
&\quad - \Omega \cdot \oint_{C_j} \mathbf{t}_j \otimes C_j\, ds_j - \Omega \cdot \oint_{C_j} C_j \otimes \mathbf{t}_j\, ds_j\\
&= -\frac{1}{2}\Omega \times \left(\oint_{C_j} C_j \times \mathbf{t}_j\, ds_j\right).\end{aligned} \tag{4.44}$$

From (4.42) and (4.44), obtain

$$\begin{aligned}\sum \oint_{C_j} \left(\frac{\delta K}{\delta C_j}\right)^\sharp (s_j)\, ds_j &= \sum \Gamma_j \oint_{C_j} \mathbf{t}_j \times \left(\frac{\partial C_j}{\partial t} + \Omega \times C_j + \mathbf{V}\right)(s_j)\, ds_j,\\ &= -\left(\frac{d}{dt} + \Omega \times\right)\left(\sum \Gamma_j \frac{1}{2}\oint_{C_j} C_j \times \mathbf{t}_j\, ds_j\right).\end{aligned}$$

Next, substitute for $(\delta K/\delta C_j)^\sharp$ in the angular momentum equation in (4.41) and resolve the integral terms using again the relations derived above.

$$
\begin{aligned}
\oint_{C_j} C_j \times \left(\frac{\delta K}{\delta C_j}\right)^\sharp (s_j)\, ds_j &= \Gamma_j \oint_{C_j} \Big((C_j \cdot \frac{\partial C_j}{\partial t})\mathbf{t}_j - (C_j \cdot \mathbf{t}_j)\frac{\partial C_j}{\partial t} \\
&\qquad - (C_j \cdot \mathbf{t}_j)(\Omega \times C_j) + C_j \times (\mathbf{t}_j \times \mathbf{V}) \Big)(s_j)\, ds_j, \\
&= \Gamma_j \oint_{C_j} \Big(\frac{1}{2}\frac{\partial \mid C_j \mid^2}{\partial t}\mathbf{t}_j + \frac{1}{2} \mid C_j \mid^2 \frac{\partial \mathbf{t}_j}{\partial t} \\
&\qquad + (\Omega \times \frac{\mid C_j \mid^2}{2}\mathbf{t}_j) - \mathbf{V} \times (C_j \times \mathbf{t}_j) - \mathbf{t}_j \times (\mathbf{V} \times C_j) \Big)(s_j)\, ds_j.
\end{aligned}
$$

The sum of the last two terms gives

$$
\begin{aligned}
&\oint_{C_j} \Big(\mathbf{V} \times (C_j \times \mathbf{t}_j) + \mathbf{t}_j \times (\mathbf{V} \times C_j) \Big)(s_j)\, ds_j \\
&= \oint_{C_j} \Big(\frac{1}{2}\mathbf{V} \times (C_j \times \mathbf{t}_j) - \frac{1}{2}((\mathbf{V}\cdot \mathbf{t}_j)C_j - \frac{1}{2}((\mathbf{V}\cdot C_j)\mathbf{t}_j) \times (\mathbf{V} \times C_j) + (\mathbf{t}_j \cdot C_j)\mathbf{V} \Big)(s_j)\, ds_j, \\
&= \mathbf{V} \times \frac{1}{2} \oint_{C_j} (C_j \times \mathbf{t}_j)(s_j)\, ds_j - \frac{\mathbf{V}}{2} \cdot \oint_{C_j} (\mathbf{t}_j \otimes C_j\, ds_j + C_j \otimes \mathbf{t}_j)(s_j)\, ds_j, \\
&= \mathbf{V} \times \frac{1}{2} \oint_{C_j} (C_j \times \mathbf{t}_j)(s_j)\, ds_j,
\end{aligned}
$$

so that finally

$$
\begin{aligned}
\oint_{C_j} C_j \times \left(\frac{\delta K}{\delta C_j}\right)^\sharp (s_j)\, ds_j &= \Gamma_j \oint_{C_j} \Big(\frac{1}{2}\frac{\partial \mid C_j \mid^2}{\partial t}\mathbf{t}_j + \frac{1}{2} \mid C_j \mid^2 \frac{\partial \mathbf{t}_j}{\partial t} \\
&\qquad + (\Omega \times \frac{\mid C_j \mid^2}{2}\mathbf{t}_j) - \mathbf{V} \times \frac{1}{2} \oint_{C_j} (C_j \times \mathbf{t}_j) \Big)(s_j)\, ds_j.
\end{aligned}
$$

Using the above relations and Eqs. (4.39)and (4.40), the linear and angular momentum equations in (4.41) can now be written as

$$
\mathbf{R} \cdot \left[\left(\frac{d}{dt} + \Omega \times\right) \left(\mathbf{L} + \sum \Gamma_j \frac{1}{2} \oint_{C_j} (C_j \times \mathbf{t}_j)(s_j)\, ds_j + \frac{1}{2} \int_{\partial B} \mathbf{l} \times (\mathbf{n} \times \mathbf{v}_\omega)\, \nu \right) ds_j \right] = 0,
$$

$$
\begin{aligned}
\mathbf{R} \cdot \Bigg[&\frac{d}{dt} \left(\hat{\mathbf{A}} - \sum \frac{1}{2} \oint_{C_j} \widehat{(\mid C_j \mid^2 \mathbf{t}_j)(s_j)\, ds_j} - \frac{1}{2} \int_{\partial B} \widehat{l^2 (\mathbf{n} \times \mathbf{v}_\omega)\, \nu} \right) \\
&+ \Omega \times \left(\mathbf{A} - \sum \frac{1}{2} \oint_{C_j} \widehat{(\mid C_j \mid^2 \mathbf{t}_j)(s_j)\, ds_j} - \frac{1}{2} \int_{\partial B} l^2 (\mathbf{n} \times \mathbf{v}_\omega)\, \nu \right) \\
&+ \mathbf{V} \times \left(\mathbf{L} + \sum \Gamma_j \frac{1}{2} \oint_{C_j} \widehat{(C_j \times \mathbf{t}_j)(s_j)\, ds_j} + \frac{1}{2} \int_{\partial B} \mathbf{l} \times (\mathbf{n} \times \mathbf{v}_\omega)\, \nu \right) \Bigg] = 0.
\end{aligned}
$$

Noting that the symbol C_j is the same as $\mathbf{l}_j$, it is easy to see that these equations are equivalent to the linear and angular momentum equations (4.27) and (4.28), respectively. And so it follows that the Euler–Lagrange equations of the system are equivalent to Eqs. (4.27)–(4.29).

4.2.2 The Vortical Momenta and Reciprocity Relations

An analysis similar to that in Sect. 3.2.2 can be carried out to obtain relations between the quantities $\mathbf{P}$ and Π and the vortical momenta integrals in the body-fixed frame, i.e. the integrals $\int_D \mathbf{v}_\omega\, \mu$ and $\int_D \mathbf{l} \times \mathbf{v}_\omega\, \mu$, respectively. Remark 2 in Sect. 1.2 should again be kept in mind while going over these relations.

These relations are presented below, and proved by explicit calculations.

Proposition 4.4. *The quantities* $\mathbf{P}$ *and* Π *defined in Eqs. (3.18) and (3.19) are related to the vortical momenta by the following relations:*

$$\mathbf{P} = \int_D \mathbf{v}_\omega\, \mu - \frac{1}{6}\sum \Gamma_j \mathbf{l}_j \times \mathbf{k} - \lim_{R\to\infty} \frac{1}{2}\int_S \mathbf{l} \times (\mathbf{n} \times \mathbf{v}_I)\ \nu,$$

$$\Pi = \int_D \mathbf{l} \times \mathbf{v}_\omega\, \mu.$$

Proof. Consider the subdomain $D_R \subset D$ with an external spherical boundary, S, of radius R (large) enclosing all the point vortices and the body and centered at the origin of the body-fixed frame. An application of (2.5) in the body-fixed frame, at a given time instant t, gives for the vortical linear momentum in D_R,

$$\int_{D_R} \mathbf{v}_\omega\, \mu = \frac{1}{2}\int_{D_R} \mathbf{l} \times \omega\, \mu + \frac{1}{2}\int_{\partial B \cup S} \mathbf{l} \times (\mathbf{n} \times \mathbf{v}_\omega)\ \nu. \qquad (4.45)$$

Now examine the behavior of the contour integral around S in the limit $R \to \infty$, $D_R \to D$. Using (4.4) and notation used previously for the image fields,write

$$\mathbf{v}_\omega(\mathbf{l}, C_1, \cdots, C_N) = \underbrace{\left[\sum \frac{\Gamma_j}{4\pi} \oint_{C_j} \frac{\mathbf{t}_j(s_j) \times (\mathbf{l}(p) - \mathbf{l}_j(s_j))}{\left|\mathbf{l}(p) - \mathbf{l}_j(s_j)\right|^3} ds_j\right]}_{\mathbf{v}_o(\mathbf{l}, C_1, \cdots, C_N)} + \mathbf{v}_I.$$

where $\mathbf{v}_I = \sum \Gamma_j \mathbf{v}^u_{I,j}(\mathbf{l})$ is the total image velocity field. Setting $\mathbf{l} = -R\mathbf{n}$, evaluate the terms in the first sum for large R,

$$\frac{\Gamma_j}{4\pi} \oint_{C_j} \frac{\mathbf{t}(s_j) \times (\mathbf{l}(p) - \mathbf{l}_j(s_j))}{\left|\mathbf{l}(p) - \mathbf{l}_j(s_j)\right|^3} ds_j$$

$$= \frac{\Gamma_j}{4\pi R^2} \oint_{C_j} \frac{\mathbf{t}_j(s_j) \times \left(\mathbf{n}(p) + \frac{\mathbf{l}_j(s_j)}{R}\right)}{\left|\mathbf{n}(p) + \frac{\mathbf{l}_j(s_j)}{R}\right|^3} ds_j,$$

$$= \frac{\Gamma_j}{4\pi R^2}\oint_{C_j} \frac{\mathbf{t}_j(s_j)\times\left(\mathbf{n}(p)+\frac{\mathbf{l}_j(s_j)}{R}\right)}{\left(1+2\frac{\mathbf{n}(p)\cdot\mathbf{l}_j(s_j)}{R}+\frac{l_j^2(s_j)}{R^2}\right)^{3/2}}\,ds_j,$$

$$= \frac{\Gamma_j}{4\pi R^2}\oint_{C_j}\left(\mathbf{t}_j(s_j)\times\left(\mathbf{n}(p)+\frac{\mathbf{l}_j(s_j)}{R}\right)\right)$$

$$\left(1-3\frac{\mathbf{n}(p)\cdot\mathbf{l}_j(s_j)}{R}+\frac{\frac{15}{4}(\mathbf{n}(p)\cdot\mathbf{l}_j(s_j))^2-\frac{3}{2}l_j^2}{R^2}+O\left(\frac{1}{R^3}\right)\right)ds_j,$$

$$= \frac{\Gamma_j}{4\pi}\oint_{C_j}\left(\frac{\mathbf{t}_j(s_j)\times\mathbf{n}(p)}{R^2}+\frac{\mathbf{t}_j(s_j)\times\mathbf{l}_j(s_j)-3(\mathbf{n}(p)\cdot\mathbf{l}_j(s_j))(\mathbf{t}(s_j)\times\mathbf{n}(p))}{R^3}+O\left(\frac{1}{R^4}\right)\right)ds_j.$$

The first term vanishes identically, and writing $\mathbf{l} = -R\mathbf{n}$ gives

$$\int_S \mathbf{l}\times(\mathbf{n}\times\mathbf{v}_o)\ \nu$$
$$= \sum\frac{\Gamma_j}{4\pi}\int_S\oint_{C_j}\left(\frac{\mathbf{t}_j(s_j)\times\mathbf{l}_j(s_j)-(\mathbf{n}(p)\otimes\mathbf{n}(p))\cdot(\mathbf{t}_j(s_j)\times\mathbf{l}_j(s_j))-3(\mathbf{n}(p)\cdot\mathbf{l}_j(s_j))(\mathbf{t}(s_j)\times\mathbf{n}(p))}{R^2}\right.$$
$$\left.+O\left(\frac{1}{R}\right)\right)ds_j\ \nu.$$

To proceed further, derive the following general relation. For any vector $\mathbf{X}(s_j)$

$$\begin{aligned}
\int_S \mathbf{n}\otimes(\mathbf{X}\times\mathbf{n})\ \nu &\equiv \int_S(\mathbf{n}\,(\mathbf{X}\times\mathbf{n})\cdot\mathbf{i},\mathbf{n}\,(\mathbf{X}\times\mathbf{n})\cdot\mathbf{j},\mathbf{n}\,(\mathbf{X}\times\mathbf{n})\cdot\mathbf{k})\ \nu,\\
&= -\frac{1}{R}\int_S(\mathbf{n}\,(\mathbf{i}\times\mathbf{X})\cdot\mathbf{l},\mathbf{n}\,(\mathbf{j}\times\mathbf{X})\cdot\mathbf{l},\mathbf{n}\,(\mathbf{k}\times\mathbf{X})\cdot\mathbf{l})\ \nu,\\
&= \frac{1}{R}\int_{D_R\cup B}(\nabla(\mathbf{i}\times\mathbf{X}\cdot\mathbf{l}),\nabla(\mathbf{j}\times\mathbf{X}\cdot\mathbf{l}),\nabla(\mathbf{k}\times\mathbf{X}\cdot\mathbf{l}))\ \mu,\\
&= \frac{1}{R}\int_{D_R\cup B}(\mathbf{i}\times\mathbf{X},\mathbf{j}\times\mathbf{X},\mathbf{k}\times\mathbf{X})\ \mu,\\
&= \frac{4\pi R^2}{3}(\mathbf{i}\times\mathbf{X},\mathbf{j}\times\mathbf{X},\mathbf{k}\times\mathbf{X}).
\end{aligned}$$

It follows that if $\mathbf{Y}(s_j)$ is another vector then,

$$\mathbf{Y}\cdot\int_S\mathbf{n}\otimes(\mathbf{X}\times\mathbf{n})\ \nu = \frac{4\pi R^2}{3}(\mathbf{X}\times\mathbf{Y}).$$

Similarly, it can be shown that $\int_S \mathbf{n}\otimes\mathbf{n}\ \nu = (4\pi R^2/3)\mathbf{Id}$, where $\mathbf{Id}$ is the 3×3 identity matrix. It follows that as $R\to\infty$,

$$\frac{1}{2}\int_S \mathbf{l}\times(\mathbf{n}\times\mathbf{v}_o)\ \nu = -\sum\frac{\Gamma_j}{6}\oint_{C_j}\mathbf{t}_j(s_j)\times\mathbf{l}_j(s_j)\,ds_j.$$

Using the far-field decay rate (4.5), it follows that $\oint_S \mathbf{l}\times(\mathbf{n}\times\mathbf{v}_I^u)\ \nu = O(1), R\to\infty$. Leaving this term in the same form, obtain as $R\to\infty$,

$$\int_D \mathbf{v}_\omega\,\mu = \frac{1}{2}\sum\Gamma_j \oint_{C_j} \mathbf{l}_j(s_j)\times\mathbf{t}_j(s_j)\,ds_j + \frac{1}{2}\oint_{\partial B} \mathbf{l}\times(\mathbf{n}\times\mathbf{v}_\omega)\ \nu$$
$$+\frac{1}{6}\sum\Gamma_j \oint_{C_j} \mathbf{l}_j(s_j)\times\mathbf{t}_j(s_j)\,ds_j + \lim_{R\to\infty}\frac{1}{2}\int_S \mathbf{l}\times(\mathbf{n}\times\mathbf{v}_I)\ \nu.$$

As in the planar case, and with reference again to Remark 2 in Sect. 1.2, $O(1)$ contributions due to the unbounded nature of the domain makes can make the variable $\mathbf{P}$ different from the linear vortical momentum integral.

The computation for the vortical angular momentum proceeds along the same lines as in the planar case. Applying (2.6) to $\mathbf{v}_\omega$ in D_R,

$$\begin{aligned}
\int_{D_R} \mathbf{l}\times\mathbf{v}_\omega\,\mu &= -\frac{1}{2}\int_{D_R} l^2\omega\,\mu - \frac{1}{2}\oint_{\partial B\cup S} l^2(\mathbf{n}\times\mathbf{v}_\omega)\ \nu,\\
&= -\frac{1}{2}\int_{D_R} l^2\omega\,\mu - \frac{1}{2}\oint_{\partial B} l^2(\mathbf{n}\times\mathbf{v}_\omega)\ \nu - \frac{R^2}{2}\oint_S (\mathbf{n}\times\mathbf{v}_\omega)\ \nu,\\
&= \Pi + \frac{R^2}{2}\oint_{D_R}\omega\,\mu + \frac{R^2}{2}\oint_B \omega\,\mu.
\end{aligned}$$

Since the domain occupied by the N rings is compact and it is implicitly assumed that there is no vorticity intersecting the body surface, an application of (4.1) shows that both the integral terms above vanish.It follows that

$$\int_D \mathbf{l}\times\mathbf{v}_\omega\,\mu = \Pi,$$

and note that this is true regardless of the value of $\sum\Gamma_j$. This proves the Proposition.

A set of reciprocity relations analogous to (3.40) exists in this problem. These were derived in [157], and are presented here. First, it was shown that there exists an analog to relation (1.17). The vector potential $\mathbf{A}_B$ associated with the Kirchhoff flow generated by the motion of the rigid body can be written relative to a body-fixed frame as:

$$\mathbf{A}_B = [A]\cdot\mathbf{U}_B, \tag{4.46}$$

where $\mathbf{U}_B = (V_1, V_2, V_3, \Omega_1, \Omega_2, \Omega_3)$ is the vector of components of the body's center-of-mass velocity and the angular velocity in the body-fixed frame's x, y, and z directions, respectively, and $[A]$ is a 3×6 matrix defined as

$$[A] := \begin{pmatrix} A_{11}\ A_{21}\ A_{31}\ A_{41}\ A_{51}\ A_{61}\\ A_{12}\ A_{22}\ A_{32}\ A_{42}\ A_{52}\ A_{62}\\ A_{13}\ A_{23}\ A_{33}\ A_{43}\ A_{53}\ A_{63}\end{pmatrix} \equiv \left(\mathbf{A}_1\ \mathbf{A}_2\ \mathbf{A}_3\ \mathbf{A}_4\ \mathbf{A}_5\ \mathbf{A}_6\right),$$

where $\nabla \times \mathbf{A}_1$ is the fluid velocity field induced by the translational motion of the body with unit speed in the x-direction, etc., and $\nabla \times \mathbf{A}_4$ is the velocity field induced by the rotational motion of the body with unit angular velocity in the x-direction, etc. These component vector potentials satisfy the boundary conditions on ∂B:

$$(\nabla \times \mathbf{A}_1)\cdot\mathbf{n} = \mathbf{i}\cdot\mathbf{n},\ \ (\nabla \times \mathbf{A}_2)\cdot\mathbf{n} = \mathbf{j}\cdot\mathbf{n},\ \ (\nabla \times \mathbf{A}_3)\cdot\mathbf{n} = \mathbf{k}\cdot\mathbf{n}, \tag{4.47}$$

$$(\nabla \times \mathbf{A}_4)\cdot\mathbf{n} = (\mathbf{i}\times\mathbf{l})\cdot\mathbf{n},\ \ (\nabla \times \mathbf{A}_5)\cdot\mathbf{n} = (\mathbf{j}\times\mathbf{l})\cdot\mathbf{n},\ \ (\nabla \times \mathbf{A}_6)\cdot\mathbf{n} = (\mathbf{k}\times\mathbf{l})\cdot\mathbf{n}, \tag{4.48}$$

so that

$$\nabla \times \mathbf{A}_B = [(\nabla \times \mathbf{A}_1)^T, (\nabla \times \mathbf{A}_2)^T, \cdots, (\nabla \times \mathbf{A}_6)^T]\cdot\mathbf{U}_B \equiv \nabla\Phi_B, \tag{4.49}$$

and it can easily be checked that the boundary condition $(\nabla \times \mathbf{A}_B)\cdot\mathbf{n} = \mathbf{U}\cdot\mathbf{n}$ is satisfied on ∂B.

It can be further shown that there exist functions $f, g, \cdots, k : \partial B \to \mathbb{R}$ such that these component vector potentials can be expressed as

$$\mathbf{A}_1 = \frac{1}{2}(\mathbf{i}\times\mathbf{l}) + \nabla_B^{\|} f,\quad \mathbf{A}_2 = \frac{1}{2}(\mathbf{j}\times\mathbf{l}) + \nabla_B^{\|} g,\quad \mathbf{A}_3 = \frac{1}{2}(\mathbf{k}\times\mathbf{l}) + \nabla_B^{\|} h,$$
$$\mathbf{A}_4 = -\frac{1}{2}l^2\mathbf{i} + \nabla_B^{\|} i,\quad \mathbf{A}_5 = -\frac{1}{2}l^2\mathbf{j} + \nabla_B^{\|} j,\quad \mathbf{A}_6 = -\frac{1}{2}l^2\mathbf{k} + \nabla_B^{\|} k.$$

These relations, and another [157], allow one to express **P**and Π as:

$$\begin{aligned}
\frac{1}{2}\int_{\partial B}\mathbf{l}\times(\mathbf{n}\times\mathbf{v}_\omega)\,\nu &= \mathbf{i}\int_{\partial B}\langle\mathbf{A}_1, \mathbf{n}\times\mathbf{v}_\omega\rangle\,\nu + \mathbf{j}\int_{\partial B}\langle\mathbf{A}_2, \mathbf{n}\times\mathbf{v}_\omega\rangle\,\nu \\
&\quad + \mathbf{k}\int_{\partial B}\langle\mathbf{A}_3, \mathbf{n}\times\mathbf{v}_\omega\rangle\,\nu, \\
-\frac{1}{2}\int_{\partial B} l^2(\mathbf{n}\times\mathbf{v}_\omega)\,\nu &= \mathbf{i}\int_{\partial B}\langle\mathbf{A}_4, \mathbf{n}\times\mathbf{v}_\omega\rangle\,\nu + \mathbf{j}\int_{\partial B}\langle\mathbf{A}_5, \mathbf{n}\times\mathbf{v}_\omega\rangle\,\nu \\
&\quad + \mathbf{k}\int_{\partial B}\langle\mathbf{A}_6, \mathbf{n}\times\mathbf{v}_\omega\rangle\,\nu.
\end{aligned} \tag{4.50}$$

Next, it is noted that the velocity fields $\nabla \times \mathbf{A}_B \equiv \nabla\Phi_B$ and $\mathbf{v}_\omega$ are L^2-orthogonal, i.e.

$$\int_D \langle\nabla \times \mathbf{A}_B, \mathbf{v}_\omega\rangle\,\mu = 0.$$

This is essentially the Hodge–Helmholtz decomposition theorem for vector fields, namely that potential fields and divergence-free fields parallel to the boundary are L^2-orthogonal.[2] An application of (2.7) then gives

[2] It can be shown by an explicit calculation, using the decay rates (4.5) and a punctured domain, that the Hodge–Helmholtz theorem still applies in this problem despite the externally unbounded nature of D and the singular distribution of vorticity.

$$\int_D \langle \mathbf{A}_B, \omega \rangle_{\mathbb{R}^3} \, \mu = - \int_{\partial B} \mathbf{A}_B \cdot \mathbf{n} \times \mathbf{v}_\omega \, \nu,$$

after showing that the boundary term on an imaginary external boundary goes to zero as the boundary goes to infinity. Substituting the vorticity distribution for rings (2.11) and using (4.50), one obtains the reciprocity relations in $\mathbb{R}^3$ as:

$$\begin{aligned} &\left(\sum \Gamma_j \oint_{C_j} \mathbf{t}_j(s_j) \cdot \mathbf{A}_1(s_j)\, ds_j, \sum \Gamma_j \oint_{C_j} \mathbf{t}_j(s_j) \cdot \mathbf{A}_2(s_j)\, ds_j, \sum \Gamma_j \oint_{C_j} \mathbf{t}_j(s_j) \cdot \mathbf{A}_3(s_j)\, ds_j \right) \\ &\qquad = -\frac{1}{2} \int_{\partial B} \mathbf{l} \times (\mathbf{n} \times \mathbf{v}_\omega)\, \nu, \\ &\left(\sum \Gamma_j \oint_{C_j} \mathbf{t}_j(s_j) \cdot \mathbf{A}_4(s_j)\, ds_j, \sum \Gamma_j \oint_{C_j} \mathbf{t}_j(s_j) \cdot \mathbf{A}_5(s_j)\, ds_j, \sum \Gamma_j \oint_{C_j} \mathbf{t}_j(s_j) \cdot \mathbf{A}_6(s_j)\, ds_j \right) \\ &\qquad = \frac{1}{2} \int_{\partial B} l^2 (\mathbf{n} \times \mathbf{v}_\omega)\, \nu. \end{aligned} \tag{4.51}$$

Using the above relations, it can be shown that (4.29) can be written as (4.32). Since H given by (4.30) differs from the Hamiltonian of the stationary rigid body, (4.16), only by the terms

$$\frac{1}{2} \left\langle (\mathbf{L}, \mathbf{A}) - (\mathbf{P}, \Pi), (\mathbf{M_b} + \mathbf{M_a})^{-1} ((\mathbf{L}, \mathbf{A}) - (\mathbf{P}, \Pi)) \right\rangle,$$

it suffices to show that the variational derivative of these terms with respect to C_j leads to the extra terms in (4.30), namely, $\mathbf{V}, \Omega \times \mathbf{l}_j$ and $\nabla \Phi_B$.

Now,

$$\begin{aligned} &\left\langle \frac{\delta}{\delta C_j} \left(\frac{1}{2} \left\langle (\mathbf{L}, \mathbf{A}) - (\mathbf{P}, \Pi), (\mathbf{M_b} + \mathbf{M_a})^{-1} ((\mathbf{L}, \mathbf{A}) - (\mathbf{P}, \Pi)) \right\rangle \right), \delta C_j \right\rangle \\ &\qquad = - \lim_{\varepsilon \to 0} \frac{1}{\varepsilon} \left[\left(\varepsilon \delta_j \mathbf{P}, \varepsilon \delta_j \Pi \right) \cdot \left(\mathbf{V}, \Omega \right) \right], \end{aligned} \tag{4.52}$$

where δ_j denotes the variation in the quantities when only C_j is varied, and

$$\begin{aligned} \delta_j \mathbf{P} = \frac{1}{2} \Gamma_j \oint_{C_j} \left(\delta \mathbf{l}_j(s_j) \times \mathbf{t}_j(s_j) + \mathbf{l}_j(s_j) \times \delta \mathbf{t}_j(s_j) \right) ds_j \\ + \delta \left(\frac{1}{2} \int_{\partial B} \mathbf{l} \times (\mathbf{n} \times \mathbf{v}_\omega)\, \nu \right). \end{aligned}$$

Recalling that $\delta \mathbf{l}_j = \delta C_j$, using the reciprocity relations (4.51) and employing previously used steps for computing variational derivatives, in particular referring to the H_a terms in Proposition 4.1, one obtains:

$$\begin{aligned}\delta_j \mathbf{P} = \Gamma_j \oint_{C_j} \Big(\delta \mathbf{l}_j(s_j) \times \mathbf{t}_j(s_j) \Big) ds_j \\ - \Gamma_j \Big(\oint_{C_j} \mathbf{t}_j \times (\nabla \times \mathbf{A}_1) \cdot \delta \mathbf{l}_j \, ds_j, \oint_{C_j} \mathbf{t}_j \times (\nabla \times \mathbf{A}_2) \cdot \delta \mathbf{l}_j \, ds_j, \\ \oint_{C_j} \mathbf{t}_j \times (\nabla \times \mathbf{A}_3) \cdot \delta \mathbf{l}_j \, ds_j \Big).\end{aligned}$$

Similarly,

$$\begin{aligned}&\delta_j \Pi \\ &= -\frac{1}{2} \left[\Gamma_j \oint_{C_j} \Big(2 \langle \delta \mathbf{l}_j, \mathbf{l}_j \rangle \mathbf{t}_j + l_j^2 \delta \mathbf{t}_j \Big) ds_j \right] - \delta \left(\frac{1}{2} \int_{\partial B} l^2 (\mathbf{n} \times \mathbf{v}_\omega) \, \nu \right), \\ &= -\Gamma_j \oint_{C_j} \Big(\langle \delta \mathbf{l}_j, \mathbf{l}_j \rangle \mathbf{t}_j - \langle \mathbf{l}_j, \mathbf{t}_j \rangle \delta \mathbf{l}_j \Big) ds_j \\ &- \Gamma_j \Big(\oint_{C_j} \mathbf{t}_j \times (\nabla \times \mathbf{A}_4) \cdot \delta \mathbf{l}_j \, ds_j, \oint_{C_j} \mathbf{t}_j \times (\nabla \times \mathbf{A}_5) \cdot \delta \mathbf{l}_j \, ds_j, \\ &\qquad \oint_{C_j} \mathbf{t}_j \times (\nabla \times \mathbf{A}_6) \cdot \delta \mathbf{l}_j \, ds_j \Big).\end{aligned}$$

Returning to (4.52) and applying some elementary vector identities and Eq. (4.49), one finally obtains

$$\begin{aligned}\left\langle \frac{\delta}{\delta C_j} \left(\frac{1}{2} \left\langle (\mathbf{L}, \mathbf{A}) - (\mathbf{P}, \Pi), (\mathbf{M_b} + \mathbf{M_a})^{-1} ((\mathbf{L}, \mathbf{A}) - (\mathbf{P}, \Pi)) \right\rangle \right), \delta C_j \right\rangle \\ = \Gamma_j \mathbf{t}_j \times (-\mathbf{V} - \Omega \times \mathbf{l}_j + \nabla \Phi_B).\end{aligned}$$

Applying the operation $(\cdots) \times \mathbf{t}_j$ gives the projection to the normal plane of C_j, which then shows that (4.29) can be written in the form (4.32).

4.3 *N* Vortex Rings Dynamically Coupled with a Rigid Sphere

As in the planar problems of the previous chapter, it is useful to consider the example where the body has a simple geometry, for example, a sphere, and also look at special configurations with additional symmetries. A schematic sketch is shown in Fig. 4.2.

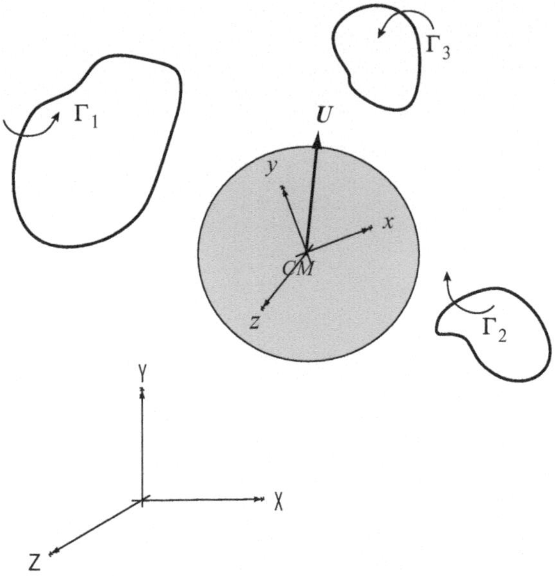

Fig. 4.2 A neutrally buoyant rigid sphere dynamically interacting with vortex rings of arbitrary strengths and signs external to it (from [158])

Previous work on sphere-vortex ring interactions has focused on a sphere that is stationary, has prescribed motion or is buoyant and is freely falling or rising. Dyson [41] briefly analyzed the passage of a circular vortex ring over a fixed sphere in an axisymmetric setting. In an important paper, Dhanak [39] carried out an analytic investigation of a sphere translating with constant speed approaching an infinitely long, initially straight vortex filament. Pedrizzetti [127] examined the motion of an straight filament in the presence of a stationary sphere in an inviscid setting and also numerically investigated the effect of adding viscous effects. This is also an important paper on the topic of secondary vortex shedding off the surface of a rigid boundary in the vicinity of a primary vortex, a topic that will be discussed in a later chapter on viscous effects. In [136], the dynamics and sound generation of an axisymmetric passage of a circular ring over a stationary sphere is investigated. Ryu and Lee [146] numerically simulated the dynamics for an axisymmetric passage and an offset passage of an initially circular ring over a stationary sphere. The latter case naturally results in a considerable distortion of the ring. Blackmore and Knio [21] presented a canonical Hamiltonian formulation of these two configurations. Dynamically coupled interactions, with buoyancy effects included, arise naturally in the problem of a rigid body falling or ascending in water. Vortical structures are sometimes generated due to viscous effects. For example, in a series of interesting papers on experiments on this problem, Horowitz and Williamson have documented the formation of vortex rings and their dynamic interaction with a falling sphere and a rising sphere [69–71].

Features analogous to previously discussed planar models where the body geometry is circular are seen in this model as well. The angular dynamics of the sphere is completely decoupled from the fluid dynamics and so, without loss of generality, one can set $\Omega = 0$. In the added mass matrix (1.8) all the ζ-terms and all terms with $\mathbf{l} \times \mathbf{n}$ are set equal to zero. It follows that, referring to (1.14), $\mathbf{M_b} + \mathbf{M_a}$ effectively reduces to the 3×3 block $\mathbf{M}_{11}$. However, a further simplification results by explicitly computing the added mass entries of $\mathbf{M}_{11}$. The expression for Φ_B for a sphere of radius a is

$$\Phi_B(\mathbf{l},t) = \sum_{i=1}^{3} V_i(t)\psi_i(\mathbf{l}),$$

where the unit potentials for a sphere are given by

$$\psi_1(\mathbf{l}) = -\frac{a^3(\mathbf{l}\cdot\mathbf{i})}{2\mid\mathbf{l}\mid^3}, \quad \psi_2(\mathbf{l}) = -\frac{a^3(\mathbf{l}\cdot\mathbf{j})}{2\mid\mathbf{l}\mid^3}, \quad \psi_3(\mathbf{l}) = -\frac{a^3(\mathbf{l}\cdot\mathbf{k})}{2\mid\mathbf{l}\mid^3}. \tag{4.53}$$

Writing $n_1 = \mathbf{n}\cdot\mathbf{i}$, etc., one obtains

$$\begin{aligned}
\int_{\partial B} \psi_1 n_1 \ \nu &= -\frac{1}{2}\int_B \nabla\cdot((\mathbf{l}\cdot\mathbf{i})\mathbf{i}) \ \mu = -\frac{2}{3}\pi a^3,\\
\int_{\partial B} \psi_2 n_1 \ \nu &= -\frac{1}{2}\int_B \nabla\cdot((\mathbf{l}\cdot\mathbf{j})\mathbf{i}) \ \mu = 0,\\
\int_{\partial B} \psi_3 n_1 \ \nu &= -\frac{1}{2}\int_B \nabla\cdot((\mathbf{l}\cdot\mathbf{k})\mathbf{i}) \ \mu = 0.
\end{aligned}$$

Similarly, the other entries leading to the classical result that the added mass of a sphere is equal to half its actual mass [92]. As in the planar case with a circle, the mass matrix therefore reduces to a scalar matrix:

$$\mathbf{M_b} + \mathbf{M_a} = \begin{pmatrix} m+\frac{2}{3}\pi a^3 & 0 & 0 \\ 0 & m+\frac{2}{3}\pi a^3 & 0 \\ 0 & 0 & m+\frac{2}{3}\pi a^3 \end{pmatrix}. \tag{4.54}$$

In addition to the conservation of the Casimirs $\mathscr{C}_1$ and $\mathscr{C}_2$, one gets from (4.27) a conservation law—stronger than $\mathscr{C}_1$—of the form given by the first equation of (1.45), i.e. $\mathbf{L}(t) = \mathbf{L}(0)$ =constant, with $\mathbf{L}$ given by (4.58). It follows from (4.58) and (4.54) that $\mathbf{A}$ is no longer a dynamical variable of the system and that H, as given by (4.30), is independent of $\mathbf{A}$. One drops equation (4.28) altogether from consideration.

Image Velocity Field

Another advantage of the sphere is that it leads to analytic expressions for the image velocity field $\mathbf{v}_I$. As one would expect, these expressions are not as simple as in Milne-Thomson's circle theorem which was used in Sect. 3.3. In particular, Knio and Ting [84] derived an expression for $\mathbf{v}_I$ using a result due to Weiss [184] on the image field in a sphere due to a potential field with singularities outside the sphere. The words "potential field" here refer to an inviscid, incompressible, and irrotational flow. The flow due to N vortex rings is an example of such a flow field and so Weiss' result applies. It is useful to present Weiss' result in the form of a theorem since it is perhaps not as popular as Milne-Thomson's result despite having greater applicability owing to its three-dimensional setting.

Theorem 4.1 (Weiss' Result [184]). *Let ϕ_0 be the velocity potential function of an inviscid, incompressible, and irrotational flow in $\mathbb{R}^3$, with possible singularities that all lie outside a spherical subdomain $S \subset \mathbb{R}^3$. If a rigid sphere is introduced in S, then the disturbance flow or image field introduced external to S is an inviscid, incompressible, and irrotational flow in $\mathbb{R}^3$, singularity-free in $\mathbb{R}^3 \backslash S$, governed by a velocity potential function ϕ_1 given by:*

$$\phi_1(\mathbf{l}) = \frac{a}{r}\phi_0\left(\frac{a^2}{l^2}\mathbf{l}\right) - \frac{2}{al}\int_0^a \lambda\phi_0\left(\frac{\lambda^2}{l^2}\mathbf{l}\right) d\lambda,$$

where a is the radius of the sphere, $\mathbf{l}$ denotes position vector from the center of S, and $l = |\mathbf{l}|$.

Weiss' work precedes and is a more general result than the work of [98] in that it is applicable to a general potential field with singularities and not just to a field of singular vortex rings or filaments. It also avoids the issue of having to deal with the structure of the image vorticity in order to obtain $\mathbf{v}_I$. It should be noted that there also exists the result due to Lewis [92, 97] which, however, applies only to axisymmetric sphere-rings configurations and is discussed more in Sect. 4.3.1. Interestingly, Lewis' work significantly predates the works of both Weiss and Milne-Thomson, and may be viewed as the generalization of the circle theorem.

Knio and Ting's formula may be written as follows. For any point q lying outside the sphere:

$$\begin{aligned} &v_I^m(q) \\ &= \sum \frac{-\Gamma_j}{4\pi a} \oint_j \left\langle \mathfrak{Q}_j(\mathbf{l},\mathbf{l}_j)\frac{\partial \mathbf{e}_l}{\partial x_m} + \left(-\mathfrak{X}_j(\mathbf{l},\mathbf{l}_j)a^2 x_m + \mathfrak{A}_j(\mathbf{l},\mathbf{l}_j)\left\langle \mathbf{e}_{l_j}, \frac{\partial \mathbf{e}_l}{\partial x_m}\right\rangle\right)\mathbf{e}_l, \mathbf{e}_{l_j} \times \mathbf{t}_j \right\rangle ds_j, \\ &\qquad m = 1,2,3 \end{aligned} \tag{4.55}$$

where $\mathbf{v}_I = (v_I^1, v_I^2, v_I^3)$, $\mathbf{l} = (x_1, x_2, x_3) \equiv (x, y, z)$ is the position vector of q in a body-fixed frame centered at the sphere. $\mathbf{e}_l$ and $\mathbf{e}_{l_j}$ denote unit radial vectors with $\mathbf{e}_l := \frac{\mathbf{l}}{|\mathbf{l}|}$ and $\mathbf{e}_{l_j} := \frac{\mathbf{l}_j}{|\mathbf{l}_j|}$, $\bar{\mathbf{l}}_j = |\bar{\mathbf{l}}_j|\, e_{l_j}$ is the position vector of the reciprocal point corresponding to the point on the ith ring with position vector $\mathbf{l}_j$, with $|\bar{\mathbf{l}}_j| = a^2/|\mathbf{l}_j|$. Note that the reciprocal points always lie inside the sphere.

The non-dimensional coefficients $\mathfrak{X}_j$, $\mathfrak{Q}_j$, and $\mathfrak{A}_j$ for each ring are defined as follows:

$$\mathfrak{Q}_j(\mathbf{l},\mathbf{l}_j) = \frac{|\bar{\mathbf{l}}|^2}{|\mathbf{l}_j|\,|\bar{\mathbf{l}}-\mathbf{l}_j|\left(\frac{|\bar{\mathbf{l}}-\mathbf{l}_j|}{|\mathbf{l}_j|}+1-\frac{1}{|\mathbf{l}_j|^2}\langle\bar{\mathbf{l}},\mathbf{l}_j\rangle\right)}$$

$$\mathfrak{X}_j(\mathbf{l},\mathbf{l}_j) = \frac{|\mathbf{l}_j|\,|\bar{\mathbf{l}}|}{|\mathbf{l}|^3\,|\bar{\mathbf{l}}-\mathbf{l}_j|^3}$$

$$\mathfrak{A}_j(\mathbf{l},\mathbf{l}_j) = \frac{|\bar{\mathbf{l}}|^3}{|\mathbf{l}_j|^2\,|\bar{\mathbf{l}}-\mathbf{l}_j|}\,\frac{\frac{|\bar{\mathbf{l}}-\mathbf{l}_j|^2}{|\mathbf{l}_j|^2}+2\frac{|\bar{\mathbf{l}}-\mathbf{l}_j|}{|\mathbf{l}_i|}+1-\frac{\langle\bar{\mathbf{l}},\mathbf{l}_j\rangle}{|\mathbf{l}_j|^2}}{\left(\frac{|\bar{\mathbf{l}}-\mathbf{l}_j|^2}{|\mathbf{l}_j|^2}+\frac{|\bar{\mathbf{l}}-\mathbf{l}_j|}{|\mathbf{l}_j|}\left(1-\frac{\langle\bar{\mathbf{l}},\mathbf{l}_j\rangle}{|\mathbf{l}_j|^2}\right)\right)^2}.$$

It is worth emphasizing that these explicit forms for the image field are only for computational purposes and are not necessary for the Hamiltonian (or Lagrangian) formulation of the system. Recall that the previous formulations for general body shapes, in $\mathbb{R}^2$ and $\mathbb{R}^3$, were obtained independently of any explicit representation of the image field.

The equations of the dynamically system are obtained from (4.27) and (4.29) as:

$$\frac{d\mathbf{L}}{dt} = 0, \tag{4.56}$$

$$\frac{\partial C_j}{\partial t} + \mathbf{V}_{|n_j} = \left(\sum_{i\neq j}\mathbf{v}_{\omega,i}+\mathbf{v}_{I,j}+\nabla\Phi_B\right)_{|n_j} + \left(\mathbf{v}_{SI,j}\right)_{|n_j}, \quad j=1,\cdots,N, \tag{4.57}$$

where

$$\mathbf{L} = (\mathbf{M_b}+\mathbf{M_a})\cdot\mathbf{V}+\mathbf{P}, \tag{4.58}$$

with $\mathbf{M_b}+\mathbf{M_a}$ given by (4.54). The vortical momentum term $\mathbf{P}$ is defined as in (4.25), but the geometry of the sphere allows one to simplify the body integral,

$$\mathbf{P} = \frac{1}{2}\sum\Gamma_j\left(\oint_{C_j}(\mathbf{l}_j(s_j)\times\mathbf{t}_j(s_j))ds_j\right) - \frac{a}{2}\int_{\partial B}\mathbf{v}_\omega dA., \tag{4.59}$$

using the boundary condition (3.13) for $\mathbf{v}_\omega$ and the convention for unit normals adopted in Chap. 1.

The Poisson bracket (4.31) of Proposition 4.2 continues to hold formally but one restricts to functions that depend only on $(\mathbf{L},C_1,\cdots,C_N)$. As in the planar problem, this leads to a trivialization of the Lie–Poisson bracket (1.38). Due to the conservation of $\mathbf{L}$ one can therefore, alternatively, restrict to a level set of $\mathbf{L}$ and view the system as a Hamiltonian system on U_{f}^N equipped with the Marsden–Weinstein Poisson bracket for rings, $\{F,G\}_{\mathrm{f}}$.

4.3.1 The Axisymmetric Model of a Sphere and N Circular Rings

The $\mathbb{R}^3$-analog of the half-space model of Sect. 3.3.1 is the axisymmetric sphere-rings model, shown in Fig. 4.3. This model may also be viewed as the extension, by the addition of the sphere, of a popular model of N-axisymmetric-vortex rings which has its roots in the nineteenth century and which merits some discussion.

The Axisymmetric N Vortex Rings Model

This model with N circular vortex rings, and no body, sharing a common central axis and lying in planes perpendicular to this axis, goes back to the pioneer of vortex dynamics—Hermann Ludwig Ferdinand Helmholtz [59]. An English translation of this paper was written by Tait [60]. This is the landmark paper in which Helmholtz postulated his famous laws of vortex motion for an inviscid, barotropic fluid. He also described in this paper the "leap-frogging" motion of two axisymmetric circular vortex rings without, however, providing analytic details. In a leap-fogging motion, one ring shrinks, accelerates, and threads through the larger, slower ring ahead of it. The lagging ring then shrinks and accelerates, and so the process repeats forever in an inviscid setting. Analytic details of this phenomenon were worked out in the papers by Gröbli [52] and Love [103], for the planar version of the problem involving pairs of counter-rotating point vortices,[3] and by Dyson [41] and Hicks for rings [63]. A good history and description of Helmholtz's paper, its massive influence on subsequent research on vortex rings—and vortex dynamics in general—combined with an exhaustive bibliography on the model (till the year 2010) may be found in the paper by Meleshko and Aref [112]. Saffman's book [147] also has a short chapter devoted to this model. It should be noted that in a lot of papers on this model, the rings are actually assumed to have a small but finite-sized circular core, with possibly time-varying core radius. The core dynamics is assumed to have negligible effect on the ring dynamics as long as it remains sufficiently small and circular.

The model continues to attract researchers, and several theoretical and numerical papers have been published in recent years that examine different features of the model. For example, Konstantinov [87] investigated chaotic phenomena in the interaction of three, four, and five rings. Blackmore and Knio applied KAM theory to study quasiperiodicity and chaos in the interaction of three rings [19, 20]; see also [18]. Blackmore, Brøns and Goullet [17] proposed a two-ring axisymmetric model with swirl to explain principal features of the real-world flow phenomenon of vortex breakdown. The instability of leap-frogging point vortex pairs to very small symmetry breaking disturbances was demonstrated numerically by Acheson [4] and later analytically, using linear Floquet theory, by Topøh and Aref [167]. The effect of an elliptic core shape was examined in [179]. Pairs of coaxial, initially elliptic rings in parallel planes have also been examined with the emphasis more on "vortex sound," i.e. the acoustics generated by a distribution of vorticity [118]. Though in such cases the rings no longer remain elliptic as they evolve, and also experience out-of-plane

[3] The half-space model without the circular cylinder.

distortions, the interesting numerical simulations by Ryu and Lee [145] show that in some cases one can still discern a mutual threading-through of the two rings akin to leap-frogging. In a real flow the symmetry about the axis is only approximate and effects like leap-frogging are eventually destroyed by viscous effects such as the diffusion of the vortex cores leading to a merging of the rings [22]. However, the configuration of axisymmetric vortex rings has also attracted the attention of experimentalists who have investigated other effects, such as the very interesting phenomenon of head-on collision of two vortex rings [100, 125].

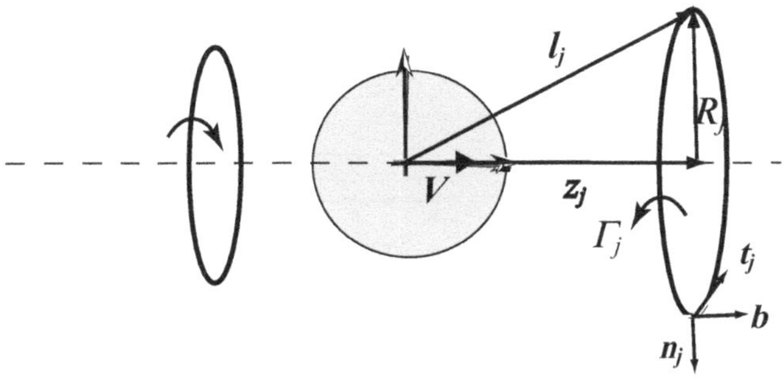

Fig. 4.3 An axisymmetric model of a neutrally buoyant rigid sphere dynamically interacting with circular rings of arbitrary strengths and signs external to it [158]

The leap-frogging motion of two rings is an example of a periodic orbit on a symmetry reduced space in the language of modern Hamiltonian systems. In each leap-frogging cycle the whole pair translates by some constant amount along the common axis of symmetry. In [155] this model was analyzed in the framework of Hamiltonian systems with symmetry, and it was shown that this translation in one cycle can be interpreted as a total reconstruction phase, i.e. the sum of a "geometric phase" and a "dynamic phase." Reconstruction phases are generically drifts in symmetry group directions that occur in Hamiltonian systems with symmetry; the geometric phase is related to geometric objects in the system's phase space such as the area enclosed by a closed orbit [107]. Note that the symmetry group in this case is identified with $\mathbb{R}$, since there is no ring evolution in the S^1-symmetry or tangential directions. In an appropriately translating frame only the periodic motion remains. In the symmetry reduced space, with appropriately chosen phase space variables, the leap-frogging motions are represented by closed curves, as in Fig. 4.4. Returning to the axisymmetric-sphere-rings model of Fig. 4.3, one can treat it as a special case of the system in Sect. 4.3 and apply the Hamiltonian formalism described there. However, it is popular in axisymmetric rings models to characterize each ring using its location along the axis and it radius. The model will therefore be described now

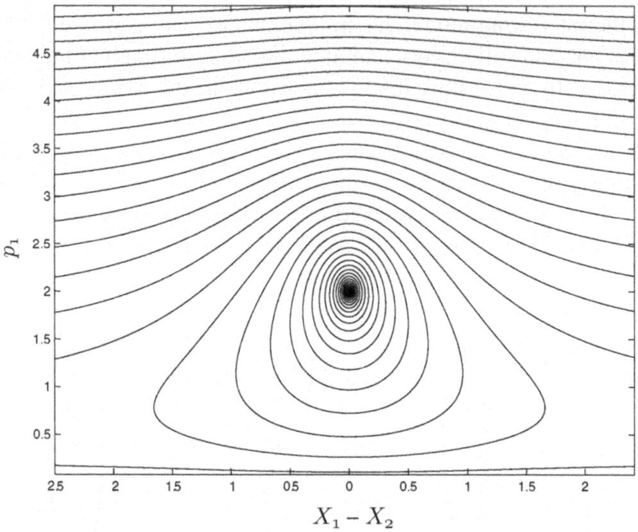

Fig. 4.4 Phase portrait of the symmetrically reduced space in the Hamiltonian model of two axisymmetric circular vortex rings. The closed curves correspond to leapfrogging motions (from [155]).

using these coordinates. Decompose the position vector of the points on the ring as follows:

$$\mathbf{l}_j(s_j) = R_j \mathbf{n}_j(s_j) \\ + z_j \mathbf{b}(s_j),$$

where (z_j, R_j) are coordinates for the jth ring, z_j is the center of the ring with respect to the center of the sphere (which is also the origin of the body-fixed frame), R_j is the radius of the ring, and $\mathbf{n}_j, \mathbf{b}$ are the unit normal and binormal vectors, respectively. Note that $\mathbf{b}$ is the same for all rings. In these coordinates, it is easily seen that the term $\mathbf{P}$ in (4.25) can be written as

$$\mathbf{P} = P\mathbf{b}, \tag{4.60}$$

$$= \left(\sum_j \Gamma_j \pi R_j^2 - \frac{a}{2} \sum_j I_j(z_j, R_j; a, \Gamma_j) \right) \mathbf{b}, \tag{4.61}$$

where a is the radius of the sphere and

$$\int_{\partial B} \mathbf{v}_{\omega,j} \, \nu =: I_j \mathbf{b}.$$

A general assumption made in the axisymmetric models, with or without the sphere, is that there is no swirl velocity or azimuthal velocity component generated. This might seem intuitively obvious, but it is useful to verify this analytically. Referring to (4.4) (after changing the subscripts to i), pick points p_i and p_j on the two rings, respectively, with parameter values s_i (varying) and s_j (fixed). Write

$$\mathbf{t}_i(s_i) = a_1(s_i)\mathbf{t}_j(s_j) + a_2(s_i)\mathbf{n}_j(s_j), \quad a_1 := \mathbf{t}_i \cdot \mathbf{t}_j,\, a_2 := \mathbf{t}_i \cdot \mathbf{n}_j,$$

$$\frac{\mathbf{r}(p_j) - \mathbf{r}(s_i)}{|\mathbf{r}(p_j) - \mathbf{r}(s_i)|^3} = \frac{1}{f(s_i)}\left(c_1(s_i)\mathbf{t}_j(s_j) + c_2(s_i)\mathbf{n}_j(s_j) + c_3\mathbf{b}\right), \quad c_3 = \text{constant}$$

$$\Rightarrow \frac{\mathbf{t}_i(s_i) \times \mathbf{r}(p_j) - \mathbf{r}(s_i)}{|\mathbf{r}(p_j) - \mathbf{r}(s_i)|^3} = \frac{1}{f}\left((a_2c_1 - a_1c_2)\mathbf{b} - a_2c_3\mathbf{t}_j + a_1c_3\mathbf{n}_j\right). \tag{4.62}$$

Using the circular geometry of the rings, one can then verify that integrating the function $a_2(s_i)/f(s_i)$ around the ith filament gives zero. Therefore, for $i \neq j$, one can write

$$\mathbf{v}_i|_{C_j} = N_{ij}(R_i, R_j, z_i, z_j; \Gamma_i)\mathbf{n}_j + B_{ij}(R_i, R_j, z_i, z_j; \Gamma_i)\mathbf{b},$$

where the N_{ij} and B_{ij} denote the coupling coefficients associated with the ith and jth ring. From (4.4), one can write these as

$$\begin{aligned} N_{ij} &= \frac{\Gamma_i}{2\pi}\int_0^{\pi} \frac{R_i(z_j - z_i)\cos\theta}{\left((z_j - z_i)^2 + R_i^2 + R_j^2 - 2R_iR_j\cos\theta\right)^{3/2}}d\theta, \\ B_{ij} &= \frac{\Gamma_i}{2\pi}\int_0^{\pi} \frac{R_i^2 - R_jR_i\cos\theta}{\left((z_j - z_i)^2 + R_i^2 + R_j^2 - 2R_iR_j\cos\theta\right)^{3/2}}d\theta. \end{aligned} \tag{4.63}$$

These integrals can be expressed in terms of the complete elliptic integrals of the first kind, $K(k) := \int_0^{\pi/2}(1 - k\sin^2\alpha)^{-1/2}d\alpha$, and of the second kind, $E(k) := \int_0^{\pi/2}(1 - k\sin^2\alpha)^{1/2}d\alpha$, as:

$$\begin{aligned} N_{ij} &= \frac{\Gamma_i R_i(z_j - z_i)}{\pi d_{ij}\sqrt{c_{ij} + d_{ij}}}\left(\frac{c_{ij}}{c_{ij} - d_{ij}}E(k_{ij}) - K(k_{ij})\right), \\ B_{ij} &= \frac{\Gamma_i}{\pi\sqrt{c_{ij} + d_{ij}}}\left(\frac{R_i^2 - R_jR_i\frac{c_{ij}}{d_{ij}}}{c_{ij} - d_{ij}}E(k_{ij}) + \frac{K(k_{ij})}{2}\right) \end{aligned} \tag{4.64}$$

where

$$c_{ij} = (z_j - z_i)^2 + R_i^2 + R_j^2, \quad d_{ij} = 2R_iR_j, \quad k_{ij} = \frac{2d_{ij}}{c_{ij} + d_{ij}}.$$

Representations in terms of elliptic integrals for the axisymmetric streamfunction and the velocity potential function in the N-axisymmetric-rings model may also be found, for example, in [40, 53, 87, 92, 147].

For the image fields, it is not obvious from Eq. (4.55) that they lack an azimuthal component. However, as mentioned in Sect. 4.3, there exists a simpler result for images in a sphere. This is due to Lewis [97] (see also the paper by Larmor [94] and the discussion by Lamb [92], art. 164) and applies specifically to axisymmetric configurations of circular rings. According to Lewis the image of an axisymmetric circular vortex ring of radius R and strength Γ located at a distance z from the center of a sphere of radius a is another axisymmetric circular vortex ring of strength whose radius, strength, and location are related to those of the original ring by the following relations:

$$z_I^2 + R_I^2 = \frac{a^4}{z^2+R^2}, \quad \frac{z_I}{R_I} = \frac{z}{R}$$
$$\Gamma_I\sqrt{R_I} = -\Gamma\sqrt{R}.$$

The first two relations are basically the same as the reciprocal relations in Milne-Thomson's circle theorem. Interestingly, the problem addressed by Lewis was that of a circular vortex ring *inside* a sphere filled with fluid but clearly the results hold regardless of whether the placement of the ring is internal or external to the sphere.

The image field is therefore another set of axisymmetric circular rings sharing the same axis. Using the same arguments as for $\mathbf{v}_i$, this allows us to conclude that for any i (including $i = j$)

$$\mathbf{v}_{I,i}{}_{|C_j} = \mathscr{N}_{ij}(R_{I,i}, R_j, z_{I,i}, z_j; \Gamma_{I,i})\mathbf{n}_j + \mathscr{B}_{ij}(R_{I,i}, R_j, z_{I,i}, z_j; \Gamma_{I,i})\mathbf{b},$$

i.e. there is no azimuthal component either. The $R_{I,i}, z_{I,i}$, and $\Gamma_{I,i}$ are obtained from R_i, z_i and Γ_i using Lewis' relations and the coefficients $\mathscr{N}_{ij}$ and $\mathscr{B}_{ij}$ can expressed in similar forms as (4.63) and (4.64) with z_i replaced by $z_{I,i}$, etc. For the Kirchhoff flow generated by the uni-directional motion of the sphere, consider, without loss of generality, only ψ_3 in (4.53). In cylindrical coordinates (r,θ,z), $\psi_3(r,\theta,z) = a^3z/2(z^2+r^2)^{3/2}$ is independent of θ and so there is no azimuthal velocity component. Therefore, write

$$\nabla\Phi_B{}_{|C_j} = (\nabla\Phi_B\cdot\mathbf{n}_j)_{|C_j}\mathbf{n}_j + (\nabla\Phi_B\cdot\mathbf{b})_{|C_j}\mathbf{b}$$

Lewis' model of image fields thus allows one to view each ring exterior to the sphere as being in the velocity field of $2N-1$ other co-axial rings in addition to its self-induced velocity field and the Kirchhoff field due to the motion of the sphere. The equations of motion for the model reduce to the following system of ODE in the variables $(L, z_1, R_1, \cdots, z_N, R_N)$:

$$\begin{aligned}
\frac{dL}{dt} &= 0,\\
\frac{dz_j}{dt} &= -\frac{1}{m+\frac{2}{3}\pi a^3}\left(L - \sum\Gamma_j\pi R_j^2 + \frac{a}{2}\sum I_j(z_j,R_j;a,\Gamma_j)\right)\\
&\quad + \sum_{i,i\neq j} B_{ij} + \sum_i \mathscr{B}_{ij} + (\nabla\Phi_B\cdot\mathbf{b})_{|C_j} + v_{\mathrm{SI},j}, \quad j=1,\cdots,N\\
\frac{dR_j}{dt} &= \sum_{i,i\neq j} N_{ij} + \sum_i \mathscr{N}_{ij} + (\nabla\Phi_B\cdot\mathbf{n}_j)_{|C_j}, \quad j=1,\cdots,N,
\end{aligned} \tag{4.65}$$

where

$$L = \left(m + \frac{2}{3}\pi a^3\right) V + \sum \Gamma_j \pi R_j^2 - \frac{a}{2} \sum I_j(z_j, R_j; a, \Gamma_j)$$

and $v_{\mathrm{SI},j}$ is the self-induced velocity assumed to be along the binormal (axis) direction. A commonly used expression for the self-induced velocity of a circular ring, again going back to the paper by Helmholtz [59]—in particular, refer to the appendix written by J. J. Thomson,—is

$$v_{\mathrm{SI},j} = \frac{\Gamma_j}{4\pi R_j} \log\left(\frac{8R_j}{r_{c,j}}\right) \mathbf{b},$$

where $r_{c,j}$ is the radius of the circular core, possibly time-varying. Compared with (2.26) this expression takes into account, to leading order, core effects. In the model, however, core effects are neglected and the magnitude of the self-induced velocity is assumed to be equal to a constant divided by the radius of the ring, i.e.

$$v_{\mathrm{SI},j} = \frac{K}{R_j}.$$

Restricting to a level set of L, one can then view the system (4.65) as residing on the phase space of the rings alone. To see what the rings bracket in (4.31) reduces to in the variables $(z_1, R_1, \cdots, z_N, R_N)$, go back to the symplectic form corresponding to the rings bracket given by Eq. (2.31). This form now acts on ring-normal vector fields. Identifying the phase space with $(\mathbb{R}^+ \times \mathbb{R})^N$, these normal vector (u, v) fields can be identified with vector fields in $\mathbb{R}^{2N}$. Let $u = \sum(r_{u,j}\mathbf{n}_j(s_j) + z_{u,j}\mathbf{b})$ and $v = \sum(r_{v,j}\mathbf{n}_j(s_j) + z_{v,j}\mathbf{b})$. Therefore,

$$\begin{aligned}
\Omega_{\mathrm{f}}(p)(u,v) &= \sum \Gamma_j \oint_{C_j} \langle \mathbf{t}_j(s_j), (r_{u,j}z_{v,j} - r_{v,j}z_{u,j})\,\mathbf{t}_j(s_j)\rangle_{\mathbb{R}^2}\, ds_j, \\
&= \sum 2\pi \Gamma_j R_j\,(r_{u,j}z_{v,j} - r_{v,j}z_{u,j}), \\
&= \sum \pi \Gamma_j dR_j^2 \wedge dz_j
\end{aligned}$$

implying that $p_j := \pi \Gamma_j R_j^2$ and z_j, $j =, 1 \cdots, N$, form a canonical set of variables for system (4.65). For the axisymmetric rings model (without the body), these coordinates were proposed by Roberts and Donnelly [137]; see also [147]. The associated Poisson brackets are

$$\{F, G\}_{\mathrm{f}} = \sum \left(\frac{\partial F}{\partial p_j}\frac{\partial G}{\partial z_j} - \frac{\partial G}{\partial p_j}\frac{\partial F}{\partial z_j}\right), \tag{4.66}$$

for $F, G : (\mathbb{R}^+ \times \mathbb{R})^N \to \mathbb{R}$.

The Case $N = 1$

For the case of only a single circular ring, all the mutual interaction terms N_{ij} and B_{ij} in (4.65), vanish. Denoting by (z_1, R_1) the coordinates of the ring with strength Γ_1, the equations reduce to

$$
\begin{aligned}
\frac{dL}{dt} &= 0, \\
\frac{dz_1}{dt} &= -\frac{1}{m + \frac{2}{3}\pi a^3}\left(L - \Gamma_1 \pi R_1^2 + \frac{a}{2} I_1(z_1, R_1; a, \Gamma_1)\right) \\
&\qquad + \mathscr{B}_{11} + (\nabla \Phi_B \cdot \mathbf{b})_{|C_1} + v_{\mathrm{SI},1}, \\
\frac{dR_1}{dt} &= \mathscr{N}_{11} + (\nabla \Phi_B \cdot \mathbf{n}_1)_{|C_1}.
\end{aligned}
\tag{4.67}
$$

An evolution for the case $L = 0, a = 1$ is shown in Fig. 4.5. The ring and sphere have initial velocities in the opposite directions. The ring expands to pass over the sphere, and during its passage also imparts momentum in the same direction to the sphere. In a real flow with viscous effects, such an interaction typically results in the generation of a secondary vortex ring off the surface of the sphere, as will be discussed more in a later chapter. It should be noted that the evolutions depicted in this figure, and also in Figures, use Weiss' equations for the image field (4.55).

It is of interest to examine if there exists an analog of the moving Föppl equilibrium of Sect. 3.3 in this axisymmetric model. This question has been examined in the classical setting of a streaming flow over a stationary sphere by Hill [64], who also endowed the ring with a non-zero core. To obtain a corresponding translating equilibrium from system (4.67), first re-introduce the sphere speed V and notice that the terms $\mathscr{B}_{11}, \mathscr{N}_{11}$, and $v_{\mathrm{SI},1}$ are linear in Γ_1 while the Kirchhoff flow terms are linear in V. Re-write the ODE system as

$$
\begin{aligned}
\frac{dz_1}{dt} &= -V + \bar{\mathscr{B}}_{11}\Gamma_1 + (\nabla \psi_1 \cdot \mathbf{b})_{|C_1} V + \bar{v}_{\mathrm{SI},1}\Gamma_1, \\
\frac{dR_1}{dt} &= \bar{\mathscr{N}}_{11}\Gamma_1 + (\nabla \psi_1 \cdot \mathbf{n}_1)_{|C_1} V,
\end{aligned}
\tag{4.68}
$$

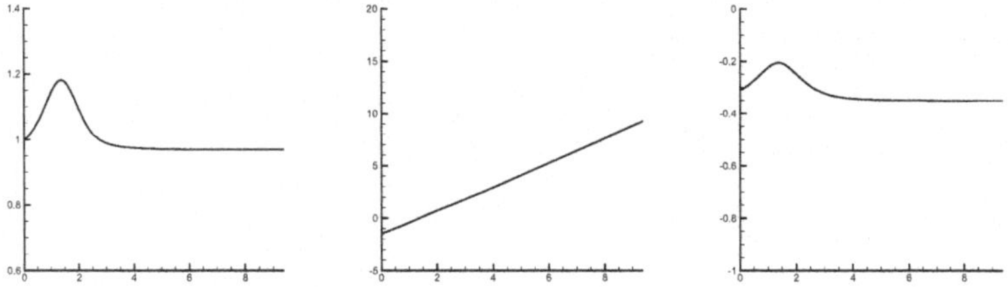

Fig. 4.5 Dynamically coupled inviscid interaction of a single circular vortex ring and a neutrally buoyant sphere. From left to right the plots show $R(t)$, $z(t)$, and $V(t)$, respectively. Numerical simulations by Mingjun Wei

where $\bar{\mathscr{B}}_{11} := \bar{\mathscr{B}}_{11}/\Gamma_1, \bar{\mathscr{N}}_{11} := \mathscr{N}_{11}/\Gamma_1$, and $\bar{v}_{\mathrm{SI},1} := v_{\mathrm{SI},1}/\Gamma_1$. The equilibrium coordinates R_1, z_1 therefore have to satisfy

$$\mathrm{Det}\begin{pmatrix} -1 - \frac{a^3(R_1^2-2z_1^2)}{2(z_1^2+R_1^2)^{5/2}} & \bar{\mathscr{B}}_{11} + \bar{v}_{\mathrm{SI},1} \\ \frac{3a^3R_1z_1}{2(z_1^2+R_1^2)^{5/2}} & \bar{\mathscr{N}}_{11} \end{pmatrix} = 0.$$

Setting $z_1 = 0$ in the above equation also generates the positions for the normal line equilibrium. All these equilibrium solutions for a sphere of unit radius and with the self-induced velocity chosen as $v_{\mathrm{SI},1} = 0.1\Gamma_1/R_1$, are plotted in Fig. 4.6. Relations between the sphere speed V and the strength of the ring Γ_1, analogous to Eq. (3.53), are then obtained as:

$$V = \Gamma_1 \frac{\bar{\mathscr{B}}_{11} + \bar{v}_{\mathrm{SI},1}}{1 + \frac{a^3(R_1^2-2z_1^2)}{2(z_1^2+R_1^2)^{5/2}}} \equiv \Gamma_1 \frac{-\bar{\mathscr{N}}_{11}}{\frac{3a^3R_1z_1}{2(z_1^2+R_1^2)^{5/2}}}.$$

Cases $N \geq 2$

A variety of different evolutions for the case of two rings is shown in Figs. 4.7 and 4.8. Two initially stationary sphere cases are first considered, Fig. 4.7, with two vortex rings of the same size at equal distances on opposite sides. In both cases, the strengths of the two rings are equal and opposite; in the bottom row the strength of each ring is reversed in sign from the top row. The sphere remains stationary and the evolutions are qualitatively similar to the classical cases in the absence of the sphere; however, there is still an image field in these two cases.

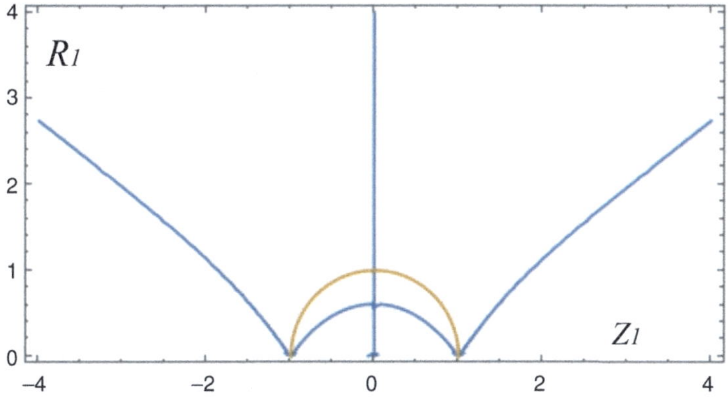

Fig. 4.6 Loci of translating equilibria ring positions in the model of a sphere and a single circular axisymmetric vortex ring. The case corresponds to a choice of $a = 1$ and $v_{\mathrm{SI},1} = 0.1\Gamma_1/R_1$. The solid gold curve represents the boundary of the sphere. External to the sphere the blue curves represent the positions of the ring in the moving Föppl equilibria, and the straight vertical line represents the positions of the ring in the moving normal line equilibrium. These equilibria are the axisymmetric analogues of the equilibria of Sect. 3.3, see Fig. 3.5

Three different cases are shown in Fig. 4.8. In all these cases, $L = 0$ and the sphere has an initial velocity along the negative z-axis. In the top row, the initially equal-sized rings with equal strengths start on opposite sides of the sphere and move in a direction opposite to the sphere. One ring passes over the sphere, and both of them approach a steadily translating evolution. In the middle row, the rings start on the same side. They both leap over the sphere and settle into a leap-frogging sequence on the other side. In the bottom row, the rings are initially of slightly unequal sizes, have opposite signs, and start on opposite sides. They both pass over the sphere and also thread through each other and approach steadily translating states in opposite directions.

In Fig. 4.9 two cases involving three rings are shown. Expectedly, the dynamics becomes more interesting and complex. For example, in the top row there are three rings of equal size initially and equal strengths. The sphere has positive initial velocity, i.e. along the positive z-axis. Two of the rings are ahead of the sphere initially and one behind the sphere. The signs of the rings strengths are such that they all move opposite to the direction of the sphere. The sphere threads through the smaller one of the pair at almost the same time as the smaller one is threading through the larger one. The pair eventually settles into a leap-frogging sequence and overtakes the third ring as well, moving in the same direction, with both of them

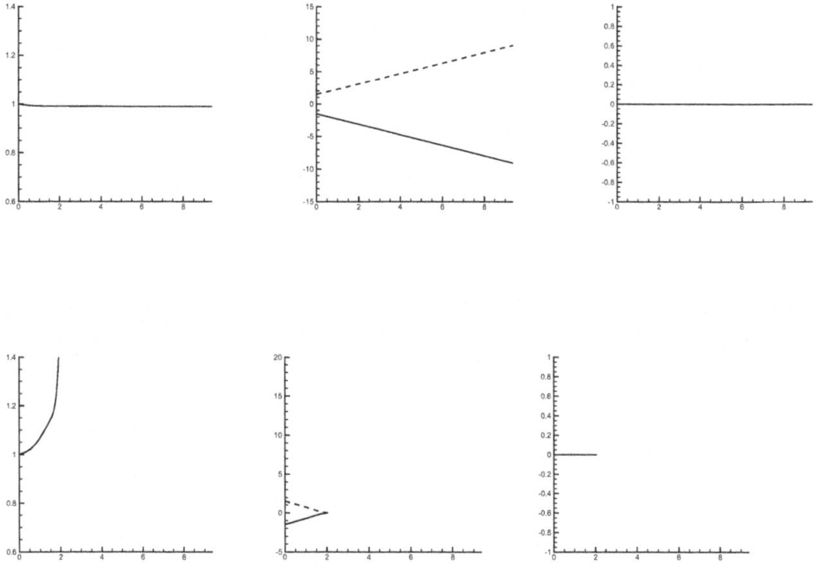

Fig. 4.7 Dynamically coupled inviscid interaction of two circular vortex rings and a neutrally buoyant sphere. From left to right the plots show $R(t)$, $z(t)$, and $V(t)$, respectively. Both rows are for cases in which $\Gamma_1 = -\Gamma_2$ and the initial conditions are $R_1(0) = R_2(0)$, $z_(0) = -z_2(0)$, and $V(0) = 0$. However, in the bottom row the strength of each ring is reversed in sign. Numerical simulations by Mingjun Wei

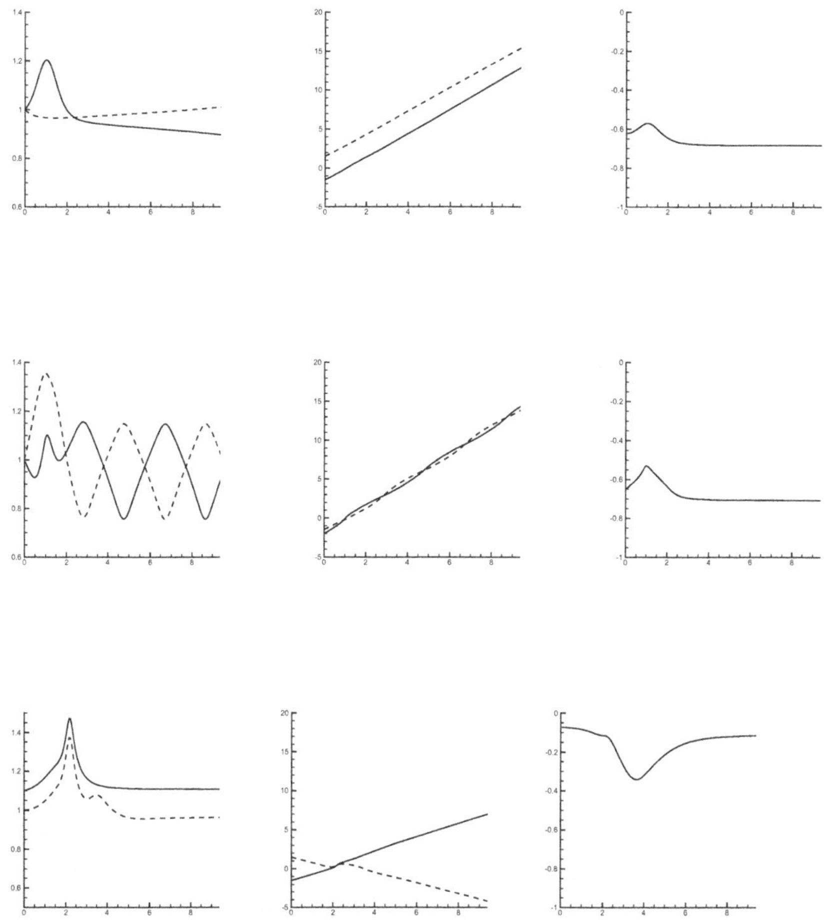

Fig. 4.8 Dynamically coupled inviscid interaction of two circular vortex rings and a neutrally buoyant sphere. From left to right the plots show $R(t)$, $z(t)$ and $V(t)$, respectively. In the top two rows, $\Gamma_1 = \Gamma_2 = 1$ and in the bottom row, $\Gamma_1 = -\Gamma_2 = 1$. The three cases are described in the text. Numerical simulations by Mingjun Wei (from [158])

threading through it. The sphere starts translating with a constant speed as it moves farther from the rings which, interestingly, is greater than its initial speed suggesting a positive transfer of momentum to the sphere from the rings.

In the bottom row, the sphere has an initial velocity in the negative z-direction. Two rings on one side of the sphere have opposite-signed strengths. The third ring on the other side passes over the sphere moving in the positive z-direction while also threading through the opposite-signed ring from the other side which moves in the negative z-direction. The third ring settles into a leap-frogging sequence with the ring of the same sign. The ring of the opposite sign eventually also overtakes the

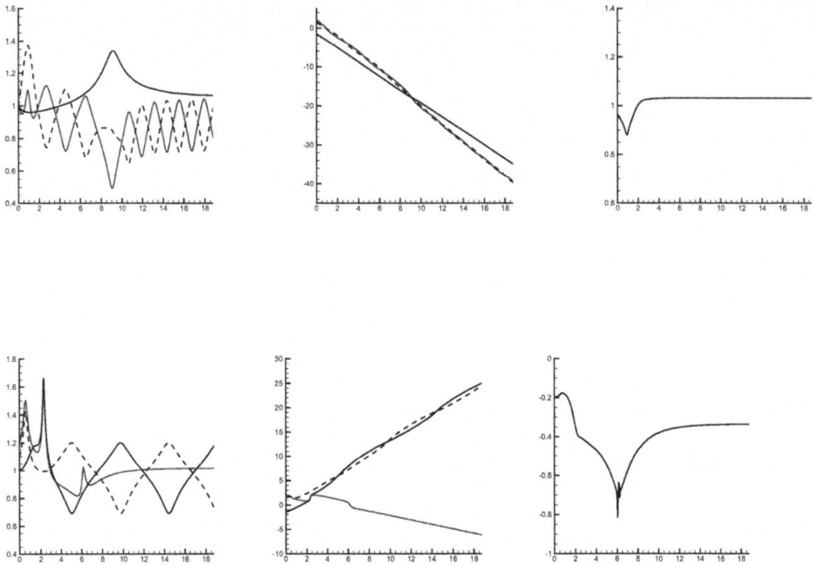

Fig. 4.9 Dynamically coupled inviscid interaction of three circular vortex rings and a neutrally buoyant sphere. From left to right the plots show $R(t)$, $z(t)$ and $V(t)$, respectively. In the top row, $\Gamma_1 = \Gamma_2 = \Gamma_3 = -1$ and in the bottom row, $\Gamma_1 = \Gamma_2 = 1$ and $\Gamma_3 = -1$ Numerical simulations by Mingjun Wei (from [158])

sphere. The sphere speed asymptotes to a constant value which, as in the previous case, is of greater magnitude than its initial speed implying a positive transfer of momentum in its original direction of travel. The greater extent of non-smoothness seen in the plots is consistent with chaotic behavior observed by previous investigators, *op. cit.*, in models of three or more rings without the body.

Chapter 5
Viscous Effects and Their Modeling

The models discussed in the previous chapters (and also the next) are all in the inviscid framework. The effects of viscosity on the coupled nonlinear dynamics of the body and fluid cannot be predicted by these models, and these effects can sometimes be important in the applications mentioned in this book. As is well known, provided the Reynolds number is sufficiently high, viscosity leads to the formation of thin boundary layers on the surface of the body. These boundary layers directly contribute to the force on the body through the mechanism of viscous shear stresses. But the indirect contribution that comes into play when the boundary layers thicken and 'separate' from the body is usually more important. Boundary layer separation considerably alters the pressure distribution on the body and, more importantly, leads to vortex shedding, i.e. the creation of new vortices off the body surface. These phenomena have been well-studied, especially in the field of aerodynamics, for steady flows. However, in these nonlinear models the flows are typically very unsteady—except at or near equilibrium configurations—and the mechanism of boundary layer separation is less well-understood for such flows. Finally the vortices themselves are subject to the effects of viscous diffusion and dissipation leading to changes in their topology, strength, and mutual interaction.

All these flow phenomena introduced due to viscosity can cause significant changes in the momentum transfer mechanism between fluid and solid. The total kinetic energy also monotonically decays in the absence of any energy-feeding mechanism into the system. The outstanding challenge is to find predictive tools that can enable these effects to be modeled.

5.1 System Momentum Balance Laws in the Viscous Setting

If the fluid domain $D \subset \mathbb{R}^3$ is externally unbounded and the rigid body assumed neutrally buoyant (with $\rho_b = \rho_f = 1$), as in the previous models, the system momenta continue to satisfy conservation laws in the absence of any external (to the whole system) forces or moments.

B. N. Shashikanth, *Dynamically Coupled Rigid Body-Fluid Flow Systems*,
https://doi.org/10.1007/978-3-030-82646-8_5

Using the vector identities (2.5) and (2.6) to represent the fluid linear and angular momentum integrals, one obtains in the domain $B \cup \widetilde{D}$, where $\tilde{D} \subset D$ is large and bounded by a sphere of radius R centered at the origin of the spatially-fixed frame,

$$\mathscr{V}_B \frac{d\mathbf{V}}{dt} + \frac{d}{dt}\left(\frac{1}{2}\left\{\int_{\tilde{D}} (\mathbf{r} \times \omega)\, \mu + \int_{\partial B} \mathbf{r} \times (\mathbf{n} \times \mathbf{u}) \nu \right\}\right) = \mathbf{F}_S$$

$$\frac{d}{dt}(\mathscr{V}_B \mathbf{r}_c \times \mathbf{V} + \mathbf{I} \cdot \Omega) - \frac{1}{2}\frac{d}{dt}\left(\int_{\tilde{D}} (r^2 \omega)\, \mu + \int_{\partial B} r^2 (\mathbf{n} \times \mathbf{u})\, \nu\right) = \mathbf{M}_S,$$

where

$$\mathbf{F}_S = \int_S \mathbf{u}(\mathbf{u} \cdot \mathbf{n})\, \nu - \frac{d}{dt}\left(\frac{1}{2}\int_S \mathbf{x} \times (\mathbf{n}_i \times \mathbf{u})\, \nu\right) + \int_S (p\mathbf{Id} - \mathbf{T}) \cdot \mathbf{n}\, \nu, \tag{5.1}$$

$$\mathbf{M}_S = \int_S \mathbf{r} \times \mathbf{u}(\mathbf{u} \cdot \mathbf{n})\, \nu + \frac{1}{2}\frac{d}{dt}\int_S r^2 (\mathbf{n} \times \mathbf{u})\, \nu + \int_S (\mathbf{r} \times ((p\mathbf{Id} - \mathbf{T}) \cdot \mathbf{n}))\, \nu. \tag{5.2}$$

In the above, $\mathbf{T} = \mu_{\text{visc}} \nabla \mathbf{u}$ denotes the deviatoric stress tensor, or the tensor of viscous stresses, for a Newtonian fluid with coefficient of viscosity μ_{visc}, p is the pressure and $\mathbf{Id}$ is the 3×3 identity matrix. The other notation is the same as in previous chapters.

Assuming the standard velocity field decay rate (4.5), which continues to hold for continuous vorticity fields, and the assumption that the vorticity field itself decays rapidly enough so that there exists a neighborhood of S in which the flow is effectively irrotational, one can show that $\mathbf{F}_S, \mathbf{M}_S \to \infty$ as $S \to \infty$ leading to the conservation laws [56][1]

$$\frac{d\mathbf{L}}{dt} = 0, \tag{5.3}$$

$$\frac{d\mathbf{A}}{dt} = 0, \tag{5.4}$$

where

$$\mathbf{L} = \mathscr{V}_B \mathbf{V} + \left(\frac{1}{2}\left(\int_D (\mathbf{r} \times \omega)\, \mu + \int_{\partial B} \mathbf{r} \times (\mathbf{n} \times (\mathbf{V} + \Omega \times \mathbf{l}))\, \nu\right)\right), \tag{5.5}$$

$$\mathbf{A} = \mathscr{V}_B \mathbf{r}_c \times \mathbf{V} + \mathbf{I} \cdot \Omega - \frac{1}{2}\left(\int_D (r^2 \omega)\, \mu + \int_{\partial B} r^2 (\mathbf{n} \times (\mathbf{V} + \Omega \times \mathbf{l}))\, \nu\right). \tag{5.6}$$

Note that in the above the no-slip boundary condition

$$\mathbf{u}_{|\partial B} = \mathbf{U}_{|\partial B}$$

has been used. Splitting the position vector as $\mathbf{r} = \mathbf{r}_c + \mathbf{l}$ and making use of the standard integral theorems and the fact that $\int_B \mathbf{l} dV = 0$, one can show that the boundary integrals can be transformed as:

[1] Details of how $\mathbf{F}_S, \mathbf{M}_S \to \infty$ in the inviscid case may be found in [157].

$$\int_{\partial B} \mathbf{r} \times (\mathbf{n} \times (\mathbf{V} + \Omega \times \mathbf{l}))\, \nu = -2(\mathbf{V} + \Omega \times \mathbf{r}_c)\mathscr{V}_B,$$
$$\int_{\partial B} r^2 (\mathbf{n} \times (\mathbf{V} + \Omega \times \mathbf{l}))\, \nu = 2r_c^2 \Omega \mathscr{V}_B + 2\Omega \int_B l^2\, \mu$$
$$+ 2(\mathbf{r}_c \times \mathbf{V})\mathscr{V}_B + 2\int_B \mathbf{l} \times (\Omega \times \mathbf{l})\, \mu.$$

It follows that

$$\mathbf{L} = \mathbf{r}_c \times \left(\Omega \mathscr{V}_B + \frac{1}{2} \int_D \omega\, \mu \right) + \frac{1}{2} \int_D \mathbf{l} \times \omega\, \mu, \tag{5.7}$$
$$\mathbf{A} = -r_c^2 \left(\Omega \mathscr{V}_B + \frac{1}{2} \int_D \omega\, \mu \right) - \frac{1}{2} \int_D l^2 \omega\, \mu - \Omega \int_B l^2\, \mu - \int_D (\mathbf{l} \cdot \mathbf{r}_c)\omega\, \mu. \tag{5.8}$$

Since $2\Omega = \nabla \times (\mathbf{V} + \Omega \times \mathbf{l})$ and the velocity field is continuous across ∂B, one can apply Stokes theorem piecewise in the body and fluid domains to show that the sum of the terms within parentheses is equal to $-(1/2)\int_S \mathbf{n} \times \mathbf{u}\, \nu$. From the decay rate, this term goes to zero as $S \to \infty$. It follows that

$$\mathbf{L} = \frac{1}{2} \int_D \mathbf{l} \times \omega\, \mu, \tag{5.9}$$
$$\mathbf{A} = -\frac{1}{2} \int_D l^2 \omega\, \mu - \Omega \int_B l^2\, \mu - \int_D (\mathbf{l} \cdot \mathbf{r}_c)\omega\, \mu, \tag{5.10}$$

and both these quantities are conserved by the coupled dynamics in a viscous interaction in $\mathbb{R}^3$. Referring to the discussion in Sect. 3.2.2, one again sees the importance of the moments of vorticity and circulation in momentum transfer between fluid and body.

The Planar Case

In $\mathbb{R}^2$, there are more conditions on the conservation of these quantities. $\mathbf{A}$ has the same form as above (with $\mathbf{I}$ reduced to a scalar, $\mathscr{V}_b$ replaced by cross-sectional area A_B) and $\mathbf{L}$ takes the form

$$\mathbf{L} = \mathbf{r}_c \times \left(2\Omega A_B + \int_D \omega\, \mu \right) + \int_D \mathbf{l} \times \omega\, \mu \tag{5.11}$$

i.e. the factor $1/2$ is no longer present. Replacing S with a circle C of radius R, the term $\mathbf{k} \cdot \int_S \mathbf{n} \times \mathbf{u}\, \nu$ now represents the circulation around C and may not go to zero as $R \to \infty$. Moreover, if $\int_D \omega\, \mu \neq 0$, then $\mathbf{u} = O(1/R)$ as $R \to \infty$ and the viscous term in $\mathbf{M}_S$ may not go to zero. The conservation of $\mathbf{A}$ in such a case is lost. For flows without boundaries, this is a result that goes back to Poincaré [130] and is also discussed in, for example [147]. To summarize, if any two of the following are always zero in a time interval of the dynamics—the circulation at infinity, $\int_D \omega\, \mu$ or Ω—then

$$\mathbf{L} = \int_D \mathbf{l} \times \omega\, \mu \tag{5.12}$$

$$\mathbf{A} = -\frac{1}{2}\int_D l^2 \omega\, \mu - \Omega \int_B l^2\, \mu - \int_D (\mathbf{l}\cdot\mathbf{r}_c)\omega\, \mu \tag{5.13}$$

are conserved by the dynamics in that interval. If $\int_D \omega\, \mu = 0$ but the circulation at infinity is not equal to zero in a time interval, then **L**, as given by (5.11), and **A** as given by (5.8) are still conserved by the dynamics. And, finally, if the circulation at infinity is equal to zero but $\int_D \omega\, \mu \neq 0$ then only **L**, as given by (5.12), is conserved.

Derivations similar in spirit, though not in a dynamically coupled setting and where the primary objective is to obtain expressions for the forces and moments exerted by the fluid on the body, may be found in, for example [72, 132, 189, 191]; see also [147]. For potential flows with isolated singularities, see [92, 117] for a discussion of classical works on force and moment expressions.

5.2 Some Experimental and Numerical Work of Vortex Rings Colliding with Rigid Bodies

In this subsection, some experimental and numerical work on a simple configuration of a dynamically coupled interaction in a viscous setting is presented. The purpose is to mainly show examples of vortex shedding due to viscous effects and also to correlate the observed dynamics with the above conservation laws.

The configuration considered also relates to another topic in body-vortex interactions: the topic of moving vortices colliding with stationary or moving objects or on near-collision paths. A few references to previous work on this topic is described in Sect. 4.3. Harvey and Perry [57] were among the first to highlight the formation and shedding of secondary vorticity off the rigid surface in the proximity of the primary vortices in their study of aircraft trailing vortices in the vicinity of the ground. The review paper by Rockwell [138] provides a good list of references to the subject of vortex–body interactions, especially focusing on collision scenarios in several areas of application. Some early papers on numerical and experimental work related to vortices colliding with stationary cylinders include, for example [33, 68, 123, 176]. Orlandi and Verzicco [124] considered the case of a vortex ring impinging on a flat wall. More recent examples are [129], which numerically considers a vortex dipole impinging on the tip of a semi-infinite plate orthogonal to the initial path of the vortex, and Guzmann et al. [54], which considers a vortex dipole colliding with a sliding wall. In this topic again, most papers typically focus on stationary objects or objects moving in a prescribed manner. The dynamically coupled setting has hardly been investigated, either numerically or experimentally.

Experimental Work

In experimental work conducted by the late James Allen of New Mexico State University, the dynamic interaction of a vortex ring approaching a neutrally buoyant rigid sphere in a water tank was examined. Full details of this experiment may be found in [6]. In Fig. 5.1 a sequence of photos illustrate the interaction. The vortex ring approaches from the left, and the sphere may be located by a light dot close to its center. Fluorescent dye and an argon-ion laser are used to visualize the vortex in a vertical plane. As any graduate student in fluid mechanics knows, flow visualization techniques depict streaklines, which could be different from streamlines or pathlines in an unsteady flow. Nevertheless, the main features of the interaction can be clearly seen in this sequence. As the primary vortex ring passes over the sphere it results in some secondary vortex shedding off the surface of the sphere and the shed vorticity in turn interacts with the primary vortex.

The interaction is more accurately captured in Fig. 5.2 which depicts velocity vector fields in the sequence of images on the left and the vorticity contour plots in the sequence of images on the right. In the latter, one clearly sees the formation of the secondary vorticity. Note also the motion of the sphere to the right under the influence of the vortex. In Fig. 5.3, the position, speed, and acceleration of the sphere versus time are plotted. Comparing the speed graph (on the right) to the speed graph for the inviscid evolution of Fig. 4.5 (rightmost graph), one sees qualitatively similar behavior before the secondary vortex shedding effects start dominating.

Numerical Work

Planar runs of the above configuration, where a symmetrically located counter-rotating vortex pair approaches a neutrally buoyant circular cylinder, were made numerically by Daniel Hartmann and Wolfgang Schröder of AIA, Germany, and are described in detail in the papers [55, 56]. A few examples of these runs are presented below. In Fig. 5.4, a time sequence of vorticity contour plots as the vortex pair approaches the circular cylinder is shown. The red colors denote positive (counter-clockwise) vorticity and the blue colors denote negative (clockwise) vorticity. In each half plane, boundary layers with opposite vorticity are seen to develop on the windward side (i.e., the side facing the approaching vortex pair) and the leeward side. The windward boundary layers separate first as the primary vortex pair approaches closer, and subsequently the leeward boundary layers. Though there is no secondary vortex formed, the secondary vorticity from these separating layers gets entrained sequentially by the primary pair. Also seen in these figures is the development of vorticity in the free shear layers along the centerline. This entire system of primary and secondary vortices and the cylinder translates to the right and, though not a moving equilibrium, it is interesting to note that it is reminiscent of the moving Föppl equilibrium of Fig. 3.6.

The numerical simulations were correlated with the computed values of the quantities (5.12) and (5.13). Note that for the symmetric case, $L_y = 0$, so that $\mathbf{L} \equiv L_x\mathbf{i} = \int_D \mathbf{l} \times \omega dA$ and since $\Omega = 0$, $\mathbf{A} = 0$. Figure 5.5, among other things,

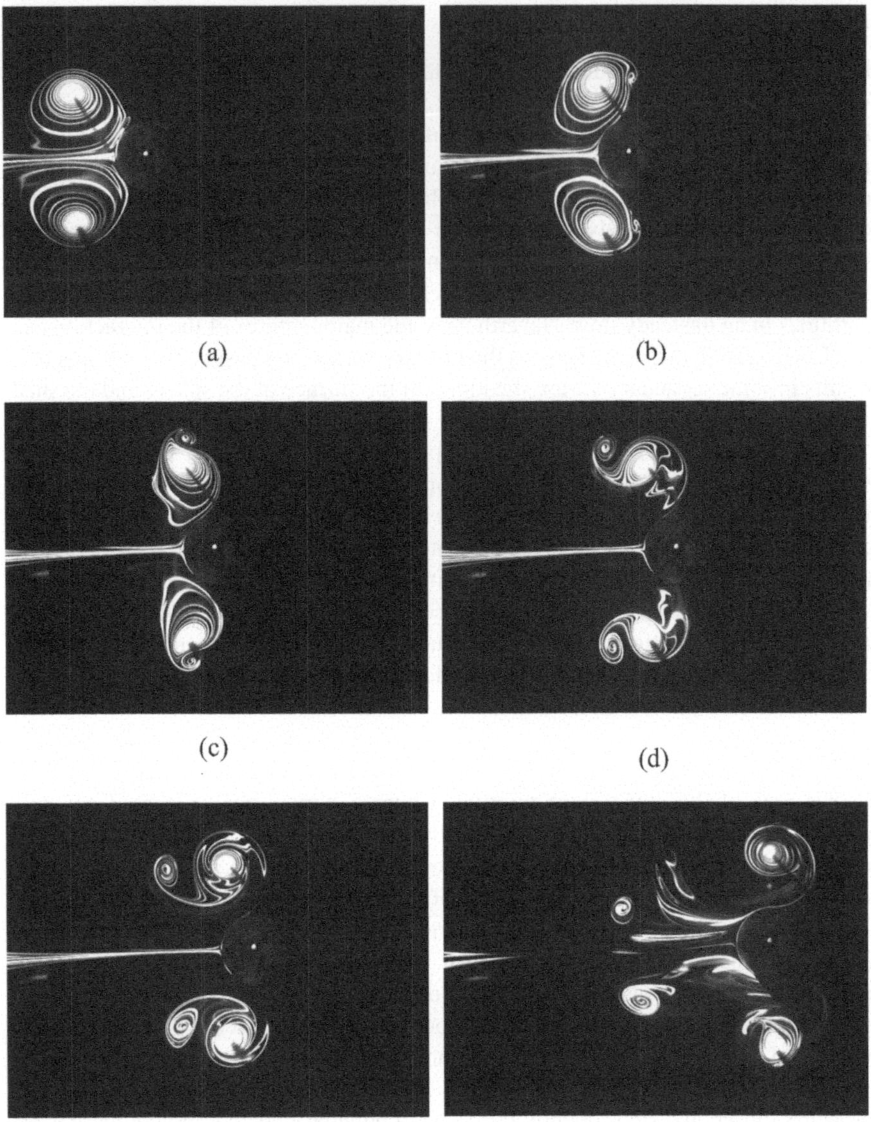

(a) (b) (c) (d) (e) (f)

Fig. 5.1 A time sequence of photographs depicting the collision of a vortex ring with a neutrally buoyant sphere. The vortex ring is visualized using fluorescent dye. Experiments by James Allen and Y. Jouanne (from [6]). ©2007 Cambridge University Press. Reprinted with permission

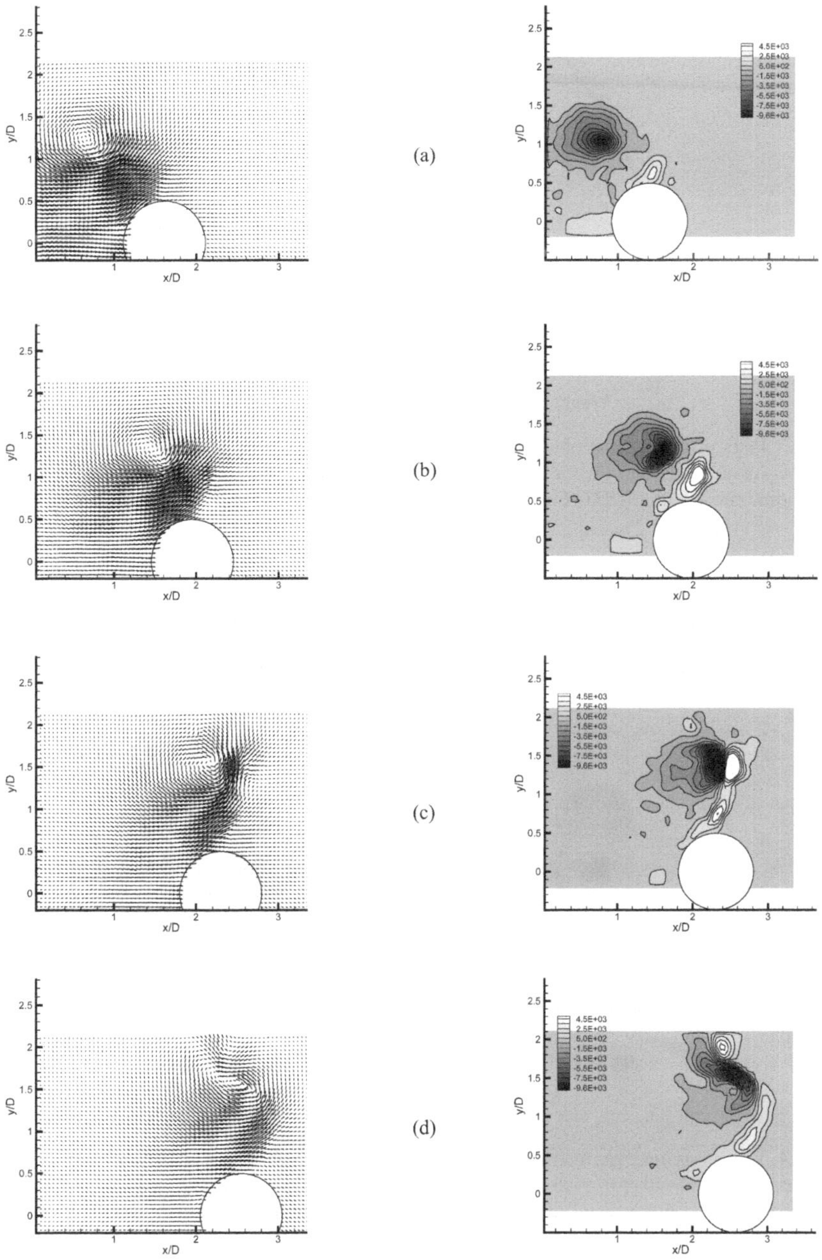

Fig. 5.2 Time evolution of velocity vector fields (left) and vorticity contour plots (right) in the interaction shown in Fig. 5.1 [6]. ©2007 Cambridge University Press. Reprinted with permission

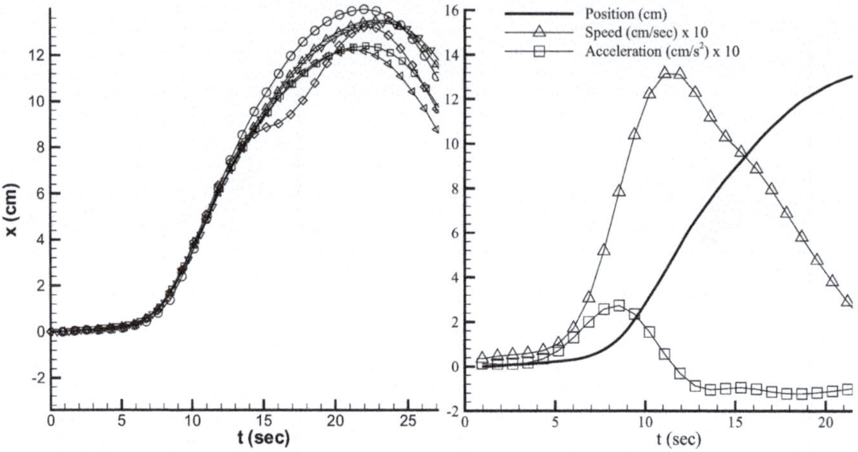

Fig. 5.3 Plots of data points obtained from the experiments. On the left is plotted the sphere's position versus time, and on the right is plotted the sphere's position, speed, and acceleration versus time (from [6]). ©2007 Cambridge University Press. Reprinted with permission

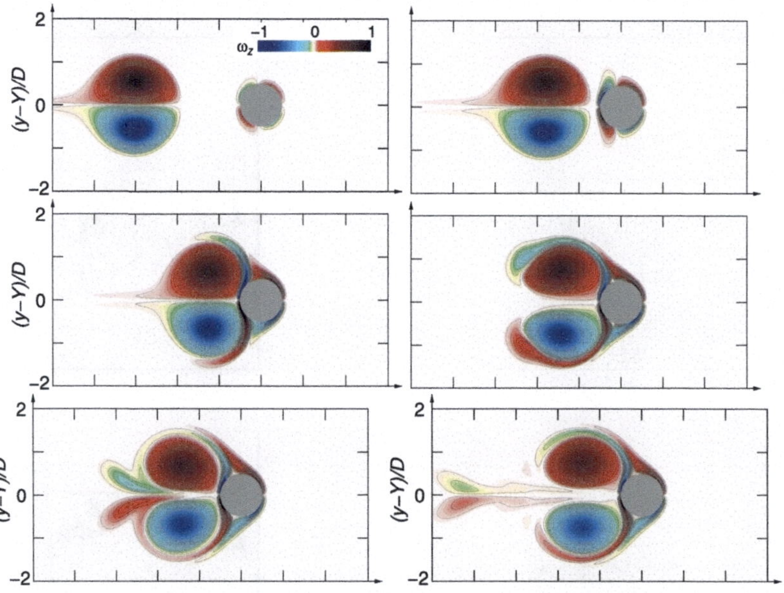

Fig. 5.4 Viscous symmetric interaction of a vortex pair approaching a neutrally buoyant circular cylinder. The x and y axes in each frame represent coordinates relative to the center of the cylinder normalized by the diameter D of the cylinder. Numerical simulations by D. Hartmann and W. Schröder. Reproduced from [56], with the permission of AIP Publishing

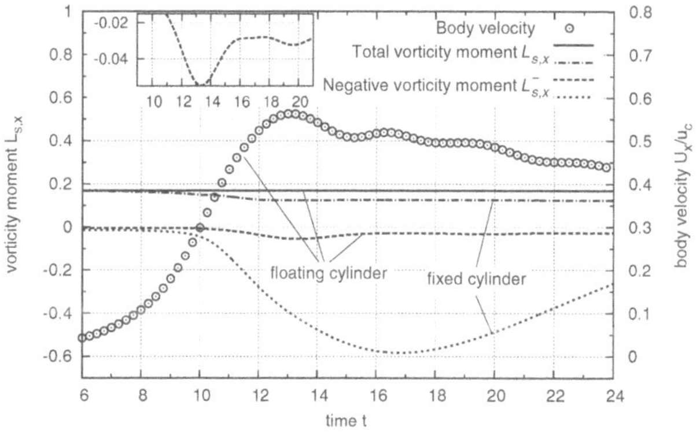

Fig. 5.5 The figure shows, among other things, correlation of behavior of **L** from simulations, depicted by the solid black line, and predicted by theory, Eq. (5.12), for the symmetric case. The solid black curve clearly shows conservation of L_x as predicted by the theory. Reproduced from [56], with the permission of AIP Publishing

shows the computed value of L_x from the simulations, over the finite computational domain, and displays excellent correlation with its behavior predicted from theory. The quantity L_x is conserved (solid black line, denoted by "$L_{s,x}$") in the moving (or "floating") cylinder case.

Simulations from an asymmetric interaction are also shown, in Fig. 5.6. Secondary vorticity formation, detachment, and merging are again observed. However, the motion of the cylinder is quite different from the symmetric case. The vortex pair overtakes the cylinder and in doing so imparts a slightly downward motion to the cylinder. This may be explained by noting that there is a separation along the x-direction in the vortex cores resulting in a generation of positive first moment of vorticity in the y-direction leading to an increase in the negative y-component of the body's linear momentum. The quantity **A** in this case is non-zero. Excellent correlation was again obtained between the behavior in the simulations and that predicted by the theory. Figure 5.7, shows among other things, the conservation of the quantity **A** of (5.13) (solid black line).

Remark

The above experiments and simulations are the results of just a few sample studies. It might be noticed in all the above pictures, for example, that the vortex core size is comparable to the cylinder size or even larger and the significant amount of secondary vortex shedding observed is partly related to this fact. Keeping the focus on these relatively simple configurations, there are many parameters in the problem that could be varied for a better understanding of the new features that viscosity introduces and their effect on the coupled dynamics and momentum and energy transfer between body and fluid: (a) initial vortex core size, (b) initial separation of vortex cores, (c) initial strengths of the vortex pair, (d) initial relative distance, and (e) ini-

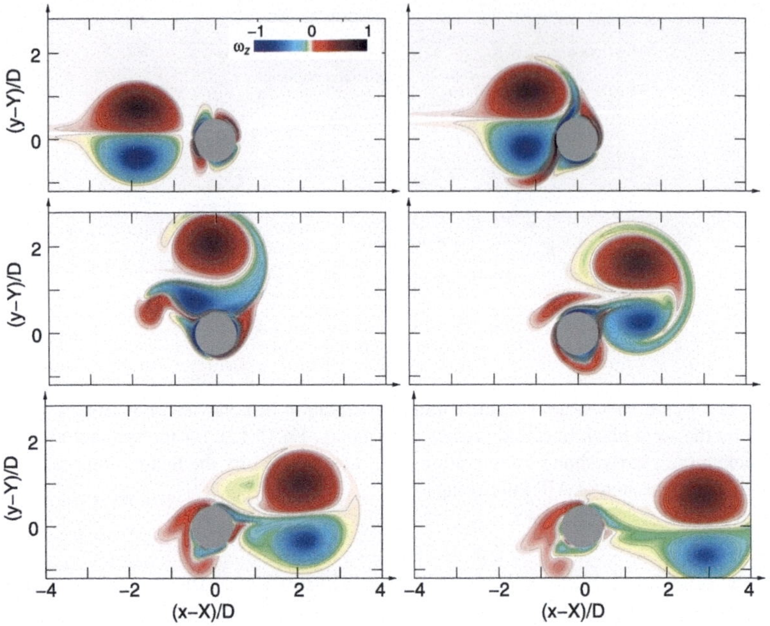

Fig. 5.6 Viscous asymmetric interaction of a vortex pair approaching a neutrally buoyant circular cylinder. The x and y axes in each frame represent coordinates relative to the center of the cylinder normalized by the diameter D of the cylinder. Numerical simulations by D. Hartmann and W. Schröder. Reproduced from [56], with the permission of AIP Publishing

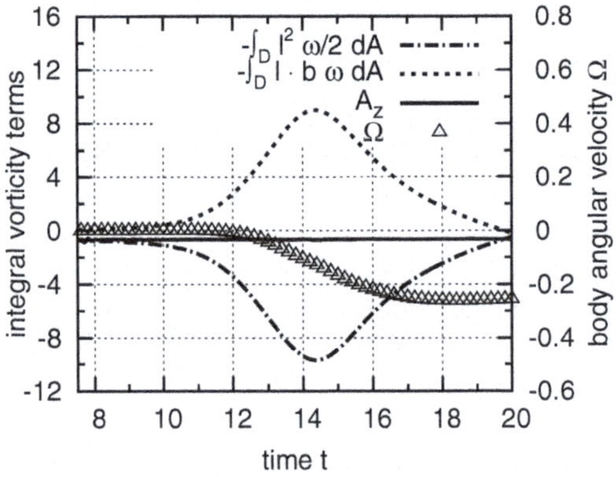

Fig. 5.7 The figure shows, among other things, correlation of behavior of **A** and that predicted by theory, Eq. (5.13). The solid black line clearly shows conservation of A_z as predicted by the theory. Reproduced from [56], with the permission of AIP Publishing

tial relative velocities of the vortex pair and the cylinder. In the asymmetric problem another parameter that could be varied is, (f) the vertical stagger of the vortex pair and the cylinder.

Inviscid Models with Vortex Shedding

Consideration of viscous flow examples like in the previous sections raises the question if the phenomenon of vortex shedding can be modeled in an inviscid framework itself, thereby approximating a principal effect of viscosity while at the same time working with the laws of inviscid fluid dynamics. The main challenge lies in predicting when and where a vortex will be shed, and this in turn is intimately related to the problem of prediction of the separation of boundary layers in unsteady flows.

This is of course a problem whose roots are in the subject of aerodynamics since vortex shedding from bodies with sharp trailing edges, such as airfoils and aircraft wings, is a commonly encountered topic in aerodynamics—even if the body is not in a dynamically coupled setting and even if flow conditions upstream are steady. For example, at large angles of attack the steady approaching flow over an aircraft wing can separate close to the leading edge itself, and this phenomenon of leading edge separation has been widely studied.

A commonly used approach to model vortex shedding is to use a vortex sheet. Saffman's book [147] contains a good introduction to vortex sheets. A vortex sheet is a codimension 1 surface supporting a singular distribution of vorticity, in contrast with point vortices and vortex rings which are codimension 2 surfaces. Geometrically, in $\mathbb{R}^2$, the vortex sheet is represented by a curve $C(s)$, and the vorticity distribution is written as

$$\omega(\mathbf{r},s) = \gamma(s)\delta(\mathbf{r}-\mathbf{r}(s))\mathbf{k},$$

where s is the curve parameter and $\gamma(s)$ is the strength density that is generally time-varying. The total strength of the sheet is $\Gamma := \int_0^l \gamma(s)\, ds$ and is a Lagrangian invariant by Kelvin's circulation theorem. The induced velocity field is again governed by the Biot–Savart law (2.4). In $\mathbb{R}^3$, the sheet is geometrically a two-dimensional surface. Across a sheet the normal component of velocity and the pressure are continuous, but there is a jump in the tangential velocity equal to the strength density at any point on the sheet. Like vortex rings and filaments, vortex sheets have self-induced velocities. In fact, a well-known phenomenon that occurs in the evolution of a vortex sheet of finite length is the roll-up at the tips, a feature that makes it a good candidate for modeling vortex shedding. Another approach is to use point vortices, with the strength of the shed vortex adjusted so as to enforce the Kutta condition at the trailing edge. Recall that the Kutta condition requires the flow to leave smoothly at the trailing edge, and in its absence inviscid fluid dynamics results in infinite velocities at sharp edges and corners.

However, most models rely on ad-hoc assumptions and apply to planar cross-sections with sharp edges. There is, to the best of the author's knowledge, still no rigorous theory to model vortex shedding in an inviscid framework in dynamically coupled problems. For a good review and history of such modeling work, see the thesis work of Michelin [113].

Chapter 6
Miscellaneous Extensions

There are other interesting dynamically coupled configurations which can also be viewed as extensions of Kirchhoff's basic model, and this chapter discusses some of them.

6.1 Dynamically Coupled Rigid Body+free Surface

The rigid body can be immersed in a semi-infinite domain of incompressible fluid externally bounded by a free surface in one vertical direction and, possibly, by impenetrable, stationary boundaries in other directions. The focus is on the dynamically coupled interaction of the body and the free surface in the presence of gravity. The other boundaries, if present, are assumed to be sufficiently far from the body to have negligible effect on its dynamics. There are, potentially, a large number of applications of this idealized model to problems in marine and ocean engineering, in particular, the applications mentioned in previous chapters.

Before discussing this extension, it is necessary to provide a brief introduction to the subject of surface water waves, i.e. the above problem without a body. This is again a topic which has its roots in the classical era but continues to attract a lot of research in modern times, much like the subject of vortices. Unlike the equations governing singular vortex motion, however, the equations governing surface water waves—neglecting vorticity and viscosity—are far more amenable to linearization techniques. Lamb's book [92] (Chapters VIII and IX on "Tidal Waves" and "Surface Waves") contains a good introduction to the topic and references to several important classical papers. The relation of the Korteweg–de Vries (KdV) equation to integrability and soliton theories has been a relatively more recent development [50] but has spawned a very large and active field of research [44]. Johnson's book [77] provides a nice introduction to both classical and modern aspects of the subject of water waves. Despite the large number of papers written on the subject, perhaps even more than on vortices, there remain many an-

B. N. Shashikanth, *Dynamically Coupled Rigid Body-Fluid Flow Systems*,
https://doi.org/10.1007/978-3-030-82646-8_6

alytic and numerical challenges due to the many complex nonlinear effects that arise in the propagation of these waves and the non-compact domains involved.

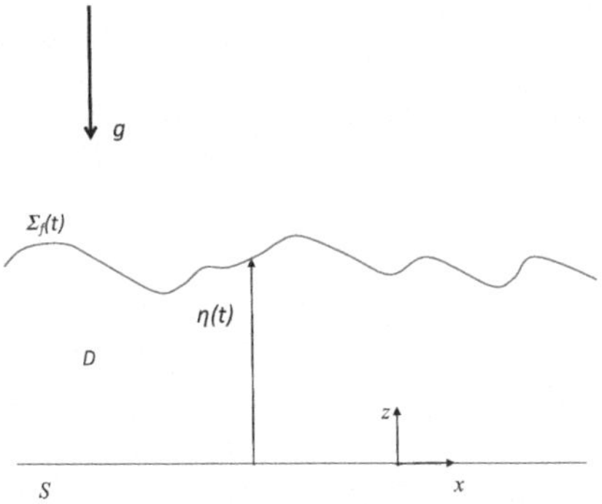

Fig. 6.1 A schematic sketch of the 3D water wave problem, showing a vertical slice with the free surface $\Sigma_f(t)$ and a flat horizontal bottom S

The surface water waves problem is typically formulated in terms of the instantaneous position of the free surface and the velocity potential function at the free surface. The air above the free surface is assumed to exert a constant pressure on the water, and the dynamics of the air is generally ignored. It is also typically assumed that the flow of water is incompressible, inviscid, and irrotational. If the dynamics of the air—or, more generally, the fluid on top—is considered important, then an alternative technique is to use vortex sheets [147]; see, for example [26]. A simple schematic sketch of a 2D slice (y=constant) in an xyz frame, is shown in Fig. 6.1. In this configuration, a flat, rigid, impenetrable bottom boundary S is assumed and there are no boundaries in the horizontal directions, but it must be added that a variety of other bottom and sidewall topography conditions have also been considered in investigations of this problem [77]. The xy-plane is parallel to S and the $z = 0$ plane–the datum–coincides with S. The non-compact fluid domain is denoted by $D \subset \mathbb{R}^{3+}$. As for vortex rings, the codimension-1 free surface can be represented as the image of maps, $\Sigma_f : \mathbb{R}^2 \to \mathbb{R}^{3+}$. A common way of representing this map is

$$\Sigma_f(x,y) \equiv (x,y,\eta(x,y)),$$

where $\eta : \mathbb{R}^2 \to \mathbb{R}$ denotes the z-coordinate of points on the surface. The convention previously adopted for vortex rings of not displaying the explicit time-dependency of these maps will also be continued here, as also the convention of using the same symbol for a map and its image. In D, the fluid flow is governed by a velocity potential function $\Phi : D \to \mathbb{R}$ satisfying

$$\nabla^2 \Phi(p) = 0, \quad p \in D \tag{6.1}$$

with Dirichlet boundary conditions at the free surface,

$$\phi(x,y,t) := \Phi(\Sigma(x,y)(t),t)_{|\Sigma_f} \equiv \Phi(x,y,\eta(x,y)(t),t)_{|\Sigma_f} = \text{prescribed at each } t, \tag{6.2}$$

and Neumann boundary conditions on the impenetrable boundaries, if present. Along unbounded directions, one imposes the standard potential flow decay conditions in $\mathbb{R}^3$, as in (4.5). It is important to remember that ϕ is viewed as an intrinsic variable for the free surface whereas Φ is a global variable defined on the spatial domain D.

As with rings, one only considers displacements of the free surface in normal directions and observes that the evolution of the free surface satisfies the following kinematic condition:

$$\frac{\partial \Sigma_f}{\partial t} = (\nabla \Phi \cdot \mathbf{n})\mathbf{n}, \tag{6.3}$$

which is the statement that the free surface is convected by the local normal velocity field and that points on the surface remain on it for all time. Viewing the surface-normal direction as a component of the vertical direction, the above condition can be expressed, alternatively, in terms of η as:

$$\frac{\partial \eta}{\partial t}\mathbf{k}\cdot\mathbf{n} = \frac{\partial \Sigma_f}{\partial t}\cdot\mathbf{n} = \nabla\Phi\cdot\mathbf{n},$$
$$\Rightarrow \frac{\partial \eta}{\partial t} = \frac{\nabla\bar{\Phi}\cdot\mathbf{n}}{\mathbf{n}\cdot\mathbf{k}}.$$

Defining $f(x,y,z) := \eta(x,y) - z$, the free surface is represented by $f = 0$ and so $\mathbf{n} = \nabla f / \mid \nabla f \mid$, and this allows the above equation to be written as:

$$\frac{\partial \eta}{\partial t} = \frac{\partial \Phi}{\partial z}_{|\Sigma_f} - \frac{\partial \Phi}{\partial x}_{|\Sigma_f}\frac{\partial \eta}{\partial x} - \frac{\partial \Phi}{\partial y}_{|\Sigma_f}\frac{\partial \eta}{\partial y}. \tag{6.4}$$

Bernoulli's equation applied at any point in D is given by

$$p = p_{atm} + \rho_f g(\eta_0 - z) - \rho_f\left(\frac{\partial \Phi}{\partial t} + \frac{1}{2}\nabla\Phi\cdot\nabla\Phi\right),$$

where η_0 is the constant value of η for the undisturbed free surface and ρ_f is the constant fluid density. In particular for points on the free surface,

$$p = p_{atm} + \rho_f g(\eta_0 - \eta) - \rho_f\left(\frac{\partial \Phi}{\partial t} + \frac{1}{2}\nabla\Phi\cdot\nabla\Phi\right)_{|\Sigma_f}.$$

The above equation can be viewed as a PDE for Φ at the free surface,

$$\frac{\partial \Phi}{\partial t}_{|\Sigma_f} = g(\eta_0 - \eta) - \frac{1}{2}(\nabla\Phi\cdot\nabla\Phi)_{|\Sigma_f} - \frac{p_{|\Sigma_f} - p_{atm}}{\rho_f}, \tag{6.5}$$

but since $\boldsymbol{\Phi}$ is defined on the entire fluid domain, it is desirable to use the intrinsic variable ϕ instead which is defined on just the free surface. From (6.2), using Σ_f or η, one obtains, respectively,

$$\frac{\partial \phi}{\partial t} = \frac{\partial \Phi}{\partial t}_{|\Sigma_f} + \nabla \Phi_{|\Sigma_f} \cdot \frac{\partial \Sigma_f}{\partial t} \equiv \frac{\partial \Phi}{\partial t}_{|\Sigma_f} + (\nabla \Phi \cdot \mathbf{n})^2$$

or

$$\frac{\partial \phi}{\partial t} = \frac{\partial \Phi}{\partial t}_{|\Sigma_f} + \nabla \Phi_{|\Sigma_f} \cdot \frac{\partial \eta}{\partial t}\mathbf{k} \equiv \frac{\partial \Phi}{\partial t}_{|\Sigma_f} + \frac{\partial \Phi}{\partial z}_{|\Sigma_f} \frac{\partial \eta}{\partial t}.$$

Bernoulli's equation then becomes

$$\frac{\partial \phi}{\partial t} = g(\eta_0 - \eta) + (\nabla \Phi \cdot \mathbf{n})^2 - \frac{1}{2}(\nabla \Phi \cdot \nabla \Phi)_{|\Sigma_f} - \frac{p_{|\Sigma_f} - p_{atm}}{\rho_f} \tag{6.6}$$

or, in conjunction with (6.4),

$$\frac{\partial \phi}{\partial t} = g(\eta_0 - \eta) + \left(\frac{\partial \Phi}{\partial z}\right)^2_{|\Sigma_f} - \frac{\partial \Phi}{\partial z}_{|\Sigma_f} \left(\frac{\partial \Phi}{\partial x}_{|\Sigma_f} \frac{\partial \eta}{\partial x} + \frac{\partial \Phi}{\partial y}_{|\Sigma_f} \frac{\partial \eta}{\partial y}\right)$$
$$- \frac{1}{2}(\nabla \Phi \cdot \nabla \Phi)_{|\Sigma_f} - \frac{p_{|\Sigma_f} - p_{atm}}{\rho_f}. \tag{6.7}$$

The pressure difference is assumed to be supported by surface tension, and the term is then replaced by the surface tension coefficient times the mean curvature of the surface at the point. In the absence of surface tension, this term is set equal to zero.

Equations (6.3) and (6.6) form a coupled set of PDE in the variables (Σ_f, ϕ). Alternatively, the more commonly used choice of the governing PDE is Eqs. (6.4) and (6.7) in the variables (η, ϕ). Since spatial derivatives of Φ appear in these equations, it would seem that one has to append equation (6.1) to close the system. The mixed Dirichlet–Neumann boundary conditions at any time instant t completely determine the solution to (6.1) at t. However, without linearizing or using other approximations, it is almost impossible to find a closed form expression for Φ in the domain D, given Σ_f and ϕ. Therefore, in numerically evolving the system one has to append equation (6.1) and solve it simultaneously. Note that whereas the Neumann boundary conditions on the impenetrable, stationary boundaries are invariant in time ($\nabla \Phi \cdot \mathbf{n} = 0$ for all t), the ϕ function in the Dirichlet condition (6.2) gets continuously updated by evolving the PDE.

The linearized wave equation is obtained by neglecting surface tension, and all quadratic derivative terms in Eqs. (6.4) and (6.5). Combining these two equations gives

$$\frac{\partial^2 \Phi}{\partial t^2}_{|\Sigma_f} + g \frac{\partial \Phi}{\partial z}_{|\Sigma_f} = 0.$$

Zakharov's Hamiltonian Formulation

Zakharov in a seminal paper [192] showed that the system of Eqs. (6.4)) and (6.7), with surface tension included, has a canonical Hamiltonian structure. The same canonical structure also holds for Eqs. (6.3)) and (6.6). Denote by $\mathscr{Z}$ the space of the (Σ_f, ϕ) variables. For the case of zero surface tension, the Hamiltonian function $H : \mathscr{Z} \to \mathbb{R}$ is the sum of the fluid kinetic and potential energy:

$$\begin{aligned} H\left(\Sigma_f, \phi\right) &= \frac{\rho_f}{2} \int_D \nabla \Phi \cdot \nabla \Phi \, \mu + \frac{1}{2} \rho_f g \left(\int_S \eta^2 \, \nu_s - \int_S \eta_0^2 \, \nu \right) \\ &= -\frac{\rho_f}{2} \int_{\Sigma_f} \Phi \nabla \Phi \cdot \mathbf{n} \, \nu + \frac{1}{2} \rho_f g \left(\int_S \eta^2 \, \nu_s - \int_S \eta_0^2 \, \nu \right), \end{aligned}$$

where ν_s denotes the area form on S, and the negative sign in the kinetic energy boundary term follows from the convention for unit normals adopted in this book. The Zakharov bracket is given by

$$\{F, K\}_{\text{Zakharov}} := \frac{1}{\rho_f} \int_{\Sigma_f} \left(\frac{\delta f}{\delta \phi} \frac{\delta k}{\delta \Sigma_f} - \frac{\delta f}{\delta \Sigma_f} \frac{\delta k}{\delta \phi} \right) \nu, \tag{6.8}$$

for functions $F, K : \mathscr{Z} \to \mathbb{R}$ of the form

$$F = \int_{\Sigma_f} f \, \nu, \tag{6.9}$$

etc.

A few remarks about the above Hamiltonian formalism, although minor, are in order: (a) Zakharov actually did not consider the rest state elevation η_0 and the corresponding rest state potential energy (which is a constant), and also assumed $\rho_f = 1$. (b) Due to the fact that Σ_f is identified with a vector field whereas ϕ is a function, to define the appropriate pairing between these dual elements and compute the variational derivatives, one formally replaces ϕ with $\phi \mathbf{n}$, a normal vector field in $\mathbb{R}^3$ with base points on Σ_f or, more precisely, by $\phi \mathbf{n}^\flat$, a normal 1-form in $\mathbb{R}^3$ with base points on Σ_f [96, 153]. Such a qualification is unnecessary when one uses Zakharov's original variables (η, ϕ). (c) The bracket above is actually the negative of the bracket Zakharov wrote, simply due to the choice of the inward pointing unit normal on Σ_f.

A detailed discussion of symmetry and conservation laws in this water wave model may be found in the paper by Benjamin and Olver [27]. An associated Lagrangian formulation for Zakharov's brackets, using the incompressibility constraint, is discussed in the review paper by Zakharov and Kuznetsov [193], as also Lagrangian and Hamiltonian formulations for a variety of other ideal flow systems.

6.1.1 A Free Surface Dynamically Coupled with a Completely Submerged Single Rigid Body of Arbitrary Shape

A rigid body is now introduced into the above setup, in proximity of the free surface but assumed to be completely submerged in the water. Like the surface water wave problem, the problem of the dynamic interactions of the surface waves with a fully or partially submerged rigid body has been well-studied; particularly the partially submerged case, some landmark publications being [75, 76, 169, 170, 181]. Indeed the papers by John and Wehausen are mostly in the dynamically coupled setting, specifically addressing the motion of floating bodies with applications to ship and marine vehicle motions, but with linearization approximations. This is typical of most approaches to this problem. Alternatively, the body is kept stationary or made to execute prescribed motions, or the objective is to derive expressions for the hydrodynamic loads on the body; see, for example [148]. Here again, like the problems of the previous chapters, the dynamically coupled *nonlinear* problem has not been as well investigated theoretically. A recent interesting theoretical paper on the subject is due to Lannes [93]. Most investigations of the nonlinear problem are typically numerical; see, for example [188] and references therein.

Papers on Lagrangian and Hamiltonian formulations of the dynamically coupled problem are even fewer. Miloh presented Lagrangian formulations in cases where the body is executing oscillations or is set in impulsive motion, on or below the free surface [114, 115]. Miloh also derived expressions for the hydrodynamical reaction forces on the body. To the best of the author's knowledge, the Hamiltonian formulation of the fully coupled system was first investigated by van Daalen, van Groesen, and Zandbergen [171, 172]; but see also [13, 14]. Working in a spatially-fixed frame and without any explicit reference to Poisson brackets, they proposed that the combined system is a canonical Hamiltonian system, with the Hamiltonian being the sum of the fluid and body kinetic+potential energies.

A schematic sketch of the setup is shown in Fig. 6.2 (only a vertical slice of the water is shown). The body boundary is again denoted by ∂B. Due to the use of a body-fixed frame, overhead bars are used for the spatially-fixed frame $\bar{x}\bar{y}\bar{z}$ (and for quantities described in it) to distinguish it from the body-fixed frame xyz, consistent with the convention used in previous chapters.

The velocity potential function $\bar{\Phi}$ in the fluid domain satisfies the following *mixed Dirichlet–Neumann* problem at each time instant t:

$$\bar{\nabla}^2\bar{\Phi} = 0 \,\text{in}\, D,\ \bar{\Phi}\,|_{\bar{\Sigma}_f} \text{ prescribed},\ \bar{\nabla}\bar{\Phi}\cdot\bar{\mathbf{n}}\,|_{\partial\bar{B}} = \bar{\mathbf{U}}\cdot\bar{\mathbf{n}},\ \bar{\nabla}\bar{\Phi}\cdot\bar{\mathbf{n}}\,|_{\bar{S}} = 0$$
$$\bar{\Phi} \to 0 \,\text{as}\, \bar{x},\bar{y} \to \pm\infty, \tag{6.10}$$

where, as in previous chapters, $\bar{\mathbf{U}} = \bar{\mathbf{V}} + \bar{\Omega} \times \bar{\mathbf{l}}$ is the rigid body velocity field in the spatially-fixed frame with $\bar{\mathbf{V}}$ denoting the center of mass velocity and $\bar{\Omega}$ its angular velocity. Both $\bar{\Phi}\,|_{\bar{\Sigma}_f}$ and $\bar{\mathbf{U}}$ are prescribed at initial time but fixed at all later times by the evolution equations.

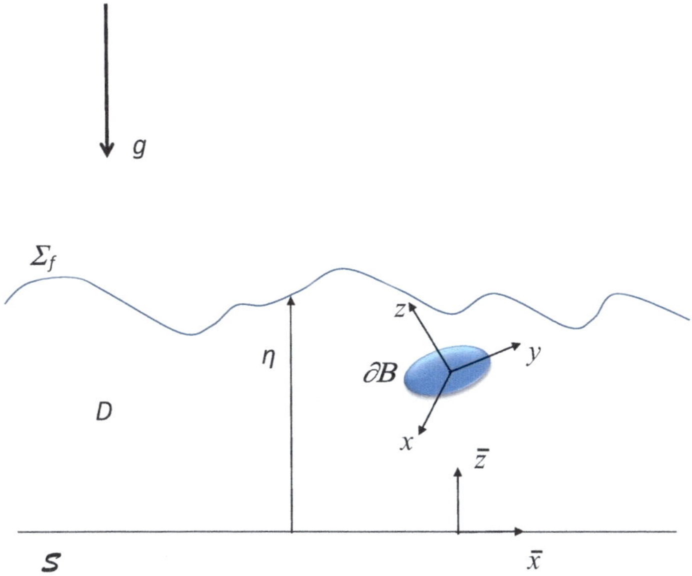

Fig. 6.2 A schematic sketch of the dynamically coupled rigid body + free surface problem in 3D. Only a 2D vertical slice of the water is shown. Also shown is the body-fixed frame (from [154])

Ignoring surface tension and surface energy, the total energy is the kinetic plus potential energy of the fluid+body system. Written in the spatially-fixed frame,

$$K.E.+P.E. = -\frac{\rho_f}{2}\left(\int_{\bar{\Sigma}_f} \bar{\Phi}\bar{\nabla}\bar{\Phi}\cdot\bar{\mathbf{n}}\,\bar{v} + \int_{\partial\bar{B}} \bar{\Phi}\bar{\nabla}\bar{\Phi}\cdot\bar{\mathbf{n}}\,\bar{v} + g\left(\int_{\bar{S}} \bar{\eta}^2\,\bar{v} - \int_{\bar{S}} \bar{\eta}_0^2\,\bar{v}\right)\right) + \frac{\rho_b}{2}\langle\langle(\mathbf{V},\Omega),\mathbf{M_b}\cdot(\mathbf{V},\Omega)\rangle\rangle_{\mathbb{R}^3} + (\rho_b-\rho_f)\mathscr{V}_B g z_c, \tag{6.11}$$

where $\mathscr{V}_B$ is the volume of the rigid body, $\mathbf{M_b}$ is the body mass tensor, and z_c is the elevation of the body centroid with respect to the datum $\mathscr{S}$. As done previously in the problem without the body, the potential energy of the fluid is written relative to the undisturbed state. This is done to subtract the infinite potential energy of the fluid in the undisturbed state. It should be noted that this can occur due to the unboundedness of the domain in horizontal directions, even if the domain is of finite extent in the vertical directions. However, subtracting the rest potential energy still does not guarantee that the fluid potential energy is finite. An additional assumption is needed about the rate at which $\bar{\eta}$ decays in the horizontal directions:

$$\int_{\bar{S}} \bar{\eta}^2\,\bar{v} - \int_{\bar{S}} \bar{\eta}_0^2\,\bar{v} < \infty. \tag{6.12}$$

Spatial Momentum Evolution

It is useful to first understand how the global body+fluid momenta in a spatially-fixed frame behave in the presence of the fixed bottom boundary and by relaxing the neutrally buoyant assumption for the body. Let

$$\bar{\mathbf{P}} \equiv (\bar{\mathbf{L}}, \bar{\mathbf{A}}) := \left(\bar{\mathbf{L}}_b + \bar{\mathbf{L}}_f, \bar{\mathbf{A}}_b + \bar{\mathbf{A}}_f\right), \tag{6.13}$$

where

$$\bar{\mathbf{L}}_b = m\bar{\mathbf{V}}, \quad \bar{\mathbf{A}}_b = \bar{\mathbf{r}}_c \times m\bar{\mathbf{V}} + \mathbf{I}\cdot\bar{\Omega} \tag{6.14}$$

are the rigid body momenta, with m the mass of the rigid body, $\bar{\mathbf{r}}_c$ the position vector of the center of mass in a spatially-fixed frame, and $\mathbf{I}$ its moment of inertia tensor in a principal-axes frame coincident with the body-fixed frame. The fluid momentum terms $\bar{\mathbf{L}}_f$ and $\bar{\mathbf{A}}_f$ are defined below.

Carrying out a traditional momentum analysis, details of which can be found in [154], one then obtains the following evolution equations for $(\bar{\mathbf{L}}, \bar{\mathbf{A}})$ in a spatially-fixed frame:

$$\frac{d\bar{\mathbf{L}}}{dt} = -\rho_f \int_{\bar{S}} \frac{\bar{\nabla}\bar{\Phi}\cdot\bar{\nabla}\bar{\Phi}}{2} \bar{\mathbf{n}}\, \bar{v} + (\rho_f - \rho_b)\, g \mathscr{V}_B \bar{\mathbf{k}}, \tag{6.15}$$

$$\frac{d\bar{\mathbf{A}}}{dt} = \int_{\bar{S}} \bar{\mathbf{r}} \times \left(-\rho_f \frac{\nabla\Phi\cdot\nabla\Phi}{2}\right) \bar{\mathbf{n}}\, \bar{v} + \bar{\mathbf{r}}_c \times (\rho_f - \rho_b)\, g \mathscr{V}_B \bar{\mathbf{k}}, \tag{6.16}$$

where

$$\bar{\mathbf{L}} = \bar{\mathbf{L}}_b \underbrace{-\rho_f \int_{\partial\bar{B}} \bar{\Phi}\bar{\mathbf{n}}\, \bar{v} - \rho_f \int_{\bar{\Sigma}_f} \bar{\Phi}\bar{\mathbf{n}}\, \bar{v}}_{\bar{\mathbf{L}}_f},$$

$$\bar{\mathbf{A}} = \bar{\mathbf{A}}_b \underbrace{-\rho_f \int_{\partial\bar{B}} \bar{\mathbf{r}} \times \bar{\Phi}\bar{\mathbf{n}}\, \bar{v} - \rho_f \int_{\bar{\Sigma}_f} \bar{\mathbf{r}} \times \bar{\Phi}\bar{\mathbf{n}}\, \bar{v}}_{\bar{\mathbf{A}}_f}$$

The following two assumptions are made in the derivation of (6.15) and (6.16). The waves on the free surface have to satisfy

$$\int_{\bar{S}} (\bar{\eta}_0 - \bar{\eta})\, \bar{v} = 0 \tag{6.17}$$

$$\int_{\bar{S}} \bar{r} \times (\bar{\eta}_0 - \bar{\eta})\, \bar{v} = 0 \tag{6.18}$$

The first relation is the global conservation of mass condition that the waves must obey. The second relation could be interpreted as zero cumulative effect of the moment about the gravity axis due to the vertical wave displacements. Since these moment vectors are actually in the horizontal plane this is a condition on zero global tilting of the total fluid mass.

It is clear that the contributions to the global momentum change come from the presence of the fixed surface $\bar{S}$ and the lack of neutral buoyancy of the rigid body. The contribution of $\bar{S}$ to the momentum change is represented by the integrals in (6.15) and (6.16). From the far-field decay rate assumptions, given by (4.5), it is easily seen that these integrals go to zero as $\bar{S}$ moves infinitely far from the free surface. To obtain global momentum conservation, one therefore needs to make the following assumptions: (a) the surface $\bar{S}$ is infinitely far from the free surface, and (b) the rigid body is neutrally buoyant. Under the above assumptions, one then obtains global momentum conservation,

$$\frac{d\bar{\mathbf{L}}}{dt} = 0 \tag{6.19}$$

$$\frac{d\bar{\mathbf{A}}}{dt} = 0. \tag{6.20}$$

For the remainder of the chapter, unless stated otherwise, it will be assumed that these conservation laws are satisfied. Further, because of assumption (b) one reverts to the setting $\rho_b = \rho_f = 1$.

Body-Fixed Frame

Equations (6.19) and (6.20) are now transformed to a body-fixed frame, with origin at the center of mass of the body and axes aligned with the principal axes, following coordinate and function transformations given by (1.20) and (1.19). The details of the transformation can again be found in [154]. Defining

$$\begin{aligned} \mathbf{L} &:= m\mathbf{V} \underbrace{- \int_{\partial B} \Phi\mathbf{n}\, \nu - \int_{\Sigma_f} \Phi\mathbf{n}\, \nu}_{\mathbf{L}_f}, \\ \mathbf{A} &:= \mathbf{I}\cdot\Omega \underbrace{- \int_{\partial B} l \times \Phi\mathbf{n}\, \nu - \int_{\Sigma_f} l \times \Phi\mathbf{n}\, \nu}_{\mathbf{A}_f}, \end{aligned} \tag{6.21}$$

the transformed equations are in Lie–Poisson form,

$$\frac{d\mathbf{L}}{dt} + \Omega \times \mathbf{L} = 0, \tag{6.22}$$

$$\frac{d\mathbf{A}}{dt} + \Omega \times \mathbf{A} + \mathbf{V} \times \mathbf{L} = 0. \tag{6.23}$$

Coupled to these are the equations for the free surface in the body-fixed frame given by

$$\frac{\partial \Sigma_f}{\partial t} = (\nabla \Phi - \mathbf{V} - \Omega \times \mathbf{l}) \cdot \mathbf{n}_{|\Sigma_f}, \tag{6.24}$$

$$\frac{\partial \phi}{\partial t} = -\left(\frac{1}{2}\nabla \Phi \cdot \nabla \Phi - (\nabla \Phi \cdot \mathbf{n})^2 + (\nabla \Phi \cdot \mathbf{n})(\mathbf{V} + \Omega \times \mathbf{l}) \cdot \mathbf{n} + g(\eta - \eta_0)\right)_{|\Sigma_f}. \tag{6.25}$$

Equations (6.22)–(6.25) form a closed system of equations governing the system. However, as in Zakharov's problem, to numerically propagate these equations the mixed Dirichlet–Neumann problem has to be appended and solved simultaneously. The mixed Dirichlet–Neumann problem in the body-fixed frame is the same as (6.10) but with the overbars removed from all variables.

Remarks

Differences may be observed in the way in which the system momenta variables are written in (6.21) and the way they are written in the vortex models—equations (3.17) and (4.58). In the latter, the affine terms are functions of the vortex variables alone and independent of the rigid body velocities, whereas in the former they are not, i.e. the integrals over the body and the free surface in (6.21) are dependent on the rigid body velocities through the Dirichlet–Neumann problem. Differences in the problem formulation may also be observed. Recall from the previous chapters, that in the vortex models it is possible to view the problem as the sum of two "component" problems: Kirchhoff's problem and the problem of the vortices interacting in the presence of an instantaneously stationary body. Such a decomposition is not possible here. For example, if one chooses as candidate component problems, Kirchhoff's problem and the problem of a free surface in the presence of an instantaneously stationary body, the addition of the Kirchhoff flow to the flow generated by the waves in the latter does not result in the flow of the dynamically coupled problem. In the vortex models, the sole effect of this addition is that the vortical structures are advected by the Kirchhoff flow (relative to the body-fixed frame) while the parameters and variables of the Neumann problem associated with the instantaneously stationary body problem, Eq. (3.1), are unaffected. Here, on the other hand, the addition also results in a change in the parameter and variables associated with the Dirichlet–Neumann problem for the instantaneously stationary body since both domain and the boundary values at the free surface are affected when the Kirchhoff flow is added. The coupling between the two problems is, in this sense, more complex.

This added complexity also influences the numerical evolution of the governing equations. Staring with an initial choice of $\mathbf{V}(0), \Omega(0), \Sigma_f(0)$, and $\phi(0)$, one obtains

$\Phi(0)$ in the fluid domain from solving the mixed Dirichlet–Neumann problem, and hence $\mathbf{L}(0)$ and $\mathbf{A}(0)$ from (6.21). The governing equations are then propagated by one time step δt which gives $\mathbf{L}(\delta t), \mathbf{A}(\delta t), \Sigma_f(\delta t)$, and $\phi(\delta t)$. Unlike in the vortex models of the previous chapter, however, obtaining $\mathbf{V}(\delta t)$ and $\Omega(\delta t)$ from (6.21) is not that straightforward due to the presence of the body surface integrals involving $\Phi(\delta t)_{|\partial B}$. The latter in turn is obtained from $\Phi(\delta t)$ in the fluid domain which requires knowledge of $\mathbf{V}(\delta t), \Omega(\delta t), \Sigma_f(\delta t)$, and $\phi(\delta t)$. An iterative scheme is therefore needed to update the values of the rigid body velocities from (6.21).

The relations (6.21) can be written in another manner in which the affine terms are independent of the rigid body velocities; however, this comes at a price. First, write

$$\Phi_{|\partial B} = \Phi^K_{|\partial B} + \Phi^{FS}_{|\partial B} + \Phi^{KFS}_{|\partial B},$$

where Φ^K is the velocity potential function of the Kirchhoff flow, Φ^{FS} the velocity potential function of the flow due to the instantaneously stationary rigid body configuration, and Φ^{KFS} is the correction that must be applied to Φ^{FS} to account for modification of the Dirichlet boundary value at the free surface due to $\Phi^K_{|\partial\Sigma}$. The integral term $\int_{\partial B} \Phi \mathbf{n} \, \nu$ is then the sum of three integrals. The integral with Φ^K can be expressed exactly like (1.7), the integral with $\Phi^{FS}_{|\partial B}$ is independent of the rigid body velocities and the integral with $\Phi^{KFS}_{|\partial B}$ can also be expressed as a symmetric mass matrix $\mathbf{M_c}$ times the rigid body velocities (see [154] for details). However, the elements of $\mathbf{M_c}$ are not constants but depend on the instantaneous shape of the free surface. Similarly, one can write

$$\Phi_{|\Sigma_f} = \Phi^K_{|\Sigma_f} + \Phi^{FS}_{|\Sigma_f},$$

where the first term can again be expressed using Kirchhoff's decomposition as a linear combination of the rigid body velocities and the second term is independent of them. But the coefficients in the first term are again dependent on the instantaneous shape of the free surface.

6.1.1.1 Phase Space and Hamiltonian Formalism

Consider the space of the $\left(\mathbf{L}, \mathbf{A}, \Sigma_f, \phi\right)$ variables, denoted by

$$\mathfrak{se}(3)^* \times \mathscr{Z}_b,$$

where $\mathscr{Z}_b$ is the space of the pairs (Σ_f, ϕ) and is isomorphic to the Zakharov space $\mathscr{Z}$ defined earlier. On this space define the Hamiltonian function $H : \mathfrak{se}(3)^* \times \mathscr{Z}_b \to \mathbb{R}$ as the total energy function (6.11) written in terms of these variables:

$$
\begin{aligned}
&H\left(\mathbf{L},\mathbf{A},\Sigma_f,\phi_f\right)\\
&=-\frac{1}{2}\left(\int_{\Sigma_f}\Phi\nabla\Phi\cdot\mathbf{n}\,\nu+\int_{\Sigma_b}\Phi\nabla\Phi\cdot\mathbf{n}\,\nu\right)+\frac{1}{2}g\left(\int_S\eta^2\,\nu-\int_S\eta_0^2\,\nu\right)\\
&\quad+\frac{1}{2}\langle\langle\mathbf{M_b}^{-1}\cdot\left[(\mathbf{L},\mathbf{A})-(\mathbf{L}_f,\mathbf{A}_f)\right],\left[(\mathbf{L},\mathbf{A})-(\mathbf{L}_f,\mathbf{A}_f)\right]\rangle\rangle_{\mathbb{R}^3}.
\end{aligned}
\tag{6.26}
$$

Now consider the following sum Poisson brackets on $\mathscr{X}_b\times\mathfrak{se}(3)^*$:

$$
\{F,G\}:=\{F|_{\mathfrak{se}(3)^*},G|_{\mathfrak{se}(3)^*}\}_{\text{Lie-Poisson}}+\{F|_{\mathscr{X}_b},G|_{\mathscr{X}_b}\}_{\text{Zakharov}}
\tag{6.27}
$$

where $\{F|_{\mathscr{X}_b},G|_{\mathscr{X}_b}\}_{\text{Zakharov}}$ is the Zakharov bracket given by (6.8) but it must be remembered now that (Σ_f,ϕ) are variables with respect to the body-fixed frame. The variational derivatives with respect to Σ_f and ϕ in (6.27) are computed as

$$
\int_{\Sigma_f}\frac{\delta f}{\delta\Sigma_f}\delta\Sigma_f\,\nu=\frac{1}{\varepsilon}\lim_{\varepsilon\to0}\left[F\left(\mathbf{L},\mathbf{A},\Sigma_f+\varepsilon\delta\Sigma_f,\phi\right)-F\left(\mathbf{L},\mathbf{A},\Sigma_f,\phi\right)\right],
\tag{6.28}
$$

$$
\begin{aligned}
\int_{\Sigma_f}\frac{\delta f}{\delta\phi}\delta\phi\,\nu\equiv\int_{\Sigma_f}\frac{\delta h}{\delta\phi}\delta\Phi\,\nu&\\
&=\frac{1}{\varepsilon}\lim_{\varepsilon\to0}\left[F\left(\mathbf{L},\mathbf{A},\Sigma_f,\phi+\varepsilon\delta\phi\right)-F\left(\mathbf{L},\mathbf{A},\Sigma_f,\phi\right)\right].
\end{aligned}
\tag{6.29}
$$

Proposition 6.1. *Equations (6.22)–(6.25) are Hamiltonian with respect to the function (6.26) and the Poisson brackets (6.27), i.e.:*

$$
\frac{\partial\Sigma_f}{\partial t}=-\frac{\delta h}{\delta\phi},
\tag{6.30}
$$

$$
\frac{\partial\phi}{\partial t}=\frac{\delta h}{\delta\Sigma_f}
\tag{6.31}
$$

$$
\frac{d(\mathbf{A},\mathbf{L})}{dt}=\mathrm{ad}^*_{\partial H/\partial(\mathbf{A},\mathbf{L})}(\mathbf{A},\mathbf{L}).
\tag{6.32}
$$

Proof. The proof consists of computing the various functional derivatives of H, and complete details may be found in [154]. An important aspect of the proof lies in showing that the variables $(\mathbf{L},\mathbf{A},\Sigma_f,\phi)$ can each be varied independently, despite the non-affine nature of relations (6.21) as discussed a few pages back.

Conserved Quantities

The presence of the Lie–Poisson bracket again means that the Casimirs $\mathscr{C}_1$and $\mathscr{C}_2$ are conserved. These conserved quantities take on the following explicit form:

$$\Big\langle m\mathbf{V}(t) - \int_{\partial B} \Phi(p_b,t)\mathbf{n}(p_b,t)\ \nu - \int_{\Sigma_f(t)} \Phi(p_f,t)\mathbf{n}(p_f,t)\ \nu,$$

$$m\mathbf{V}(t) - \int_{\partial B} \Phi(p_b,t)\mathbf{n}(p_b,t)\ \nu - \int_{\Sigma_f(t)} \Phi(p_f,t)\mathbf{n}(p_f,t)\ \nu \Big\rangle_{\mathbb{R}^3} = \text{constant},$$

$$\Big\langle m\mathbf{V}(t) - \int_{\partial B} \Phi(p_b,t)\mathbf{n}(p_b,t)\ \nu - \int_{\Sigma_f(t)} \Phi(p_f,t)\mathbf{n}(p_f,t)\ \nu,$$

$$\mathbf{I}\cdot\Omega(t) - \int_{\partial B} l(p_b)\times\Phi(p_b,t)\mathbf{n}(p_b,t)\ \nu - \int_{\Sigma_f(t)} l(p_f)\times\Phi(p_f,t)\mathbf{n}(p_f,t)\ \nu \Big\rangle_{\mathbb{R}^3}$$

$$= \text{constant},$$

where $p_b \in \partial B$ and $p_f \in \Sigma_f$.

The Kirchhoff and Zakharov Limiting Cases If the free surface is moved to infinity and toward an unperturbed state, then using

$$\Phi_{|\Sigma_f} \to 0, \quad \eta \to \eta_0,$$

the following are easily verified: Eq. (6.25) goes to zero, Eq. (6.24) goes to the equation for the rigid free surface referred to the body-fixed frame, the free surface integral terms in (6.21) go to zero, and the momenta variables reduce to those in Kirchhoff's problem. Equations (6.22) and (6.23) thus reduce to Kirchhoff's equations of motion. All terms of the form (6.9) are zero in (6.26), and so the Lie algebra of functions for the Zakharov bracket is trivial. The Zakharov bracket vanishes from (6.27) and one is left with just the Lie–Poisson bracket and the Hamiltonian kinetic energy of Kirchhoff's problem.

Similarly, removing the rigid body and the body-fixed frame by setting equal to zero the mass matrix, all integrals over Σ_b and the $(\mathbf{V},\Omega)$ terms, one obtains the Zakharov system (in the absence of surface tension). Equation (6.32), in this case, reduces to the conservation laws (6.19) and (6.20).

6.2 Dynamically Coupled N Rigid Bodies in the Absence of Vorticity

Another interesting extension involves introducing other rigid bodies in the flow and considering their dynamically coupled interaction, while retaining all the other assumptions in Kirchhoff's model. Consider N such neutrally buoyant rigid bodies of arbitrary (smooth) shape, as shown in Fig. 6.3.

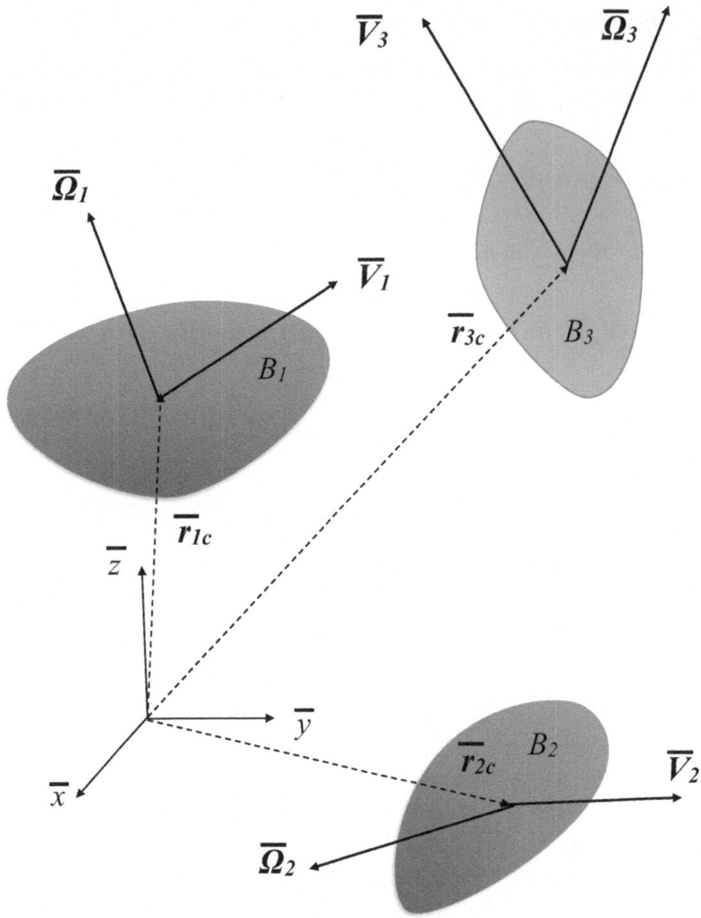

Fig. 6.3 A schematic sketch of three dynamically coupled neutrally buoyant rigid bodies in an inviscid, incompressible, and irrotational flow

Abdel Moneim [1] analyzed the problem for general rigid shapes using the Kirchhoff decomposition into unit potentials and also computed expressions for the forces and torques on the various bodies; see also [178]. An analysis in a geometric mechanics framework for general rigid shapes was carried out by Kanso, Marsden, Rowley, and Melli-Huber [79] in which the theory of Lagrangian reduction of stages developed in [32] was applied. Details were worked out for the case $N = 3$. These theories were applied directly in a body-fixed frame attached to an arbitrarily chosen body; see also Cendra and Marsden [31] where these Lagrangian reduction theories are applied to the problem of relative motion between two asteroids under mutual gravitational attraction.

A special case of this hydrodynamic "N-body problem" is the interaction of two spheres. This is another problem with roots going back to the nineteenth century. In these classical works again the focus is not on the dynamically coupled interaction but rather on obtaining approximations for the velocity potential function when the spheres have prescribed motions or studying the effect of small prescribed oscillations of the spheres on each other. Lamb [92] discusses this problem and the classical papers. In more recent times, the N-sphere model has been extended and studied with a different application in view—the problem of interacting bubbles of spherical shape but with varying volume; see, for example [74, 177]. Recall from the models of the previous chapters that for spherical/circular shapes the inviscid flow assumption implies that the rotational degree of freedom is ineffective, and so in these bubble models the dynamic interplay is between the translational and dilatational degrees of freedom. It may be noted that the problem of multiple rigid spheres in a fluid at the other end of the Reynolds number spectrum, i.e. at low Reynolds numbers, has also been investigated building on Stokes' famous results for a single sphere.

In the following, the Euler–Lagrange equations of the system are presented in a spatially-fixed frame. This does not require the Kirchhoff decomposition into unit potentials.

6.2.1 The Euler–Lagrange Equations in a Spatially-Fixed Frame

In a spatially-fixed frame, the velocity potential of the flow $\bar{\Phi}$ satisfies the Neumann problem:

$$\nabla^2\bar{\Phi} = 0, \quad \bar{\nabla}\bar{\Phi}\cdot\bar{\mathbf{n}}_i = \bar{\mathbf{U}}_i\cdot\bar{\mathbf{n}}_i \text{ on } \partial\bar{B}_i, \quad i = 1,\cdots,N, \tag{6.33}$$

where $\bar{\mathbf{U}}_i = \bar{\mathbf{V}}_i + \bar{\Omega}_i \times (\bar{\mathbf{r}} - \bar{\mathbf{r}}_{ci})_{|\partial B_i}$. For each body one has the following relation:

$$\bar{\mathbf{r}} = \mathbf{R}_i\cdot(\mathbf{l}_i + \mathbf{r}_{ic}) = \bar{\mathbf{l}}_i + \bar{\mathbf{r}}_{ic}, \quad i = 1,\cdots,N,$$

where $\bar{\mathbf{r}}_{ic}$ locates the ith body's center of mass in the spatially-fixed frame, $\mathbf{R}_i$ is the rotation matrix corresponding to the orientation of the body-fixed frame attached to the ith body relative to the spatially-fixed frame, and $\bar{\mathbf{l}}_i$ is the position vector from the center of mass of the ith body. Unlike in Kirchhoff's problem, $\bar{\Phi}$ is dependent not only on the shape of the bodies and their velocities, but also on their positions and their orientations. Due to the SE(3) symmetry in the problem, however, it is reasonable to assume that this configurational dependency appears only through the relative positions of the centers of mass and the relative orientations. This implies that the variables $\bar{\mathbf{r}}_{jc}, \mathbf{R}_j$, etc. ($j = 1,\cdots,N$) occur only in the combinations $\bar{\mathbf{r}}_{jc} - \bar{\mathbf{r}}_{ic}$, $(\mathbf{R}_j - \mathbf{R}_i)\cdot\bar{\mathbf{X}}$, etc.

The total kinetic energy of the fluid is

$$
\begin{aligned}
2T_f &= \int_{\bar{D}} \bar{\nabla}\bar{\Phi} \cdot \bar{\nabla}\bar{\Phi}\, \bar{\mu}, \\
&= -\sum_{j=1}^{N} \int_{\partial \bar{B}_j} \bar{\Phi}\bar{\nabla}\bar{\Phi} \cdot \bar{\mathbf{n}}_j\, \bar{v}, \\
&= -\sum_{j=1}^{N} \left\langle \left(\int_{\partial \bar{B}_j} \bar{\Phi}\bar{\mathbf{n}}_j\, \bar{v}, \int_{\partial \bar{B}_j} \bar{\Phi}(\bar{\mathbf{l}}_j \times \bar{\mathbf{n}}_j) \right) \cdot (\bar{\mathbf{V}}_j, \bar{\Omega}_j) \right\rangle .
\end{aligned}
$$

Consider the Lagrangian $\mathscr{L} : (T\,\mathrm{SE}(3))^N \to \mathbb{R}$ as $T_f + T_b$ viewed as a function of the variables on $(T\,\mathrm{SE}(3))^N$ and written as

$$
\begin{aligned}
&\mathscr{L}\left(\mathbf{R}_1, \cdots, \mathbf{R}_N, \bar{\mathbf{r}}_{1c}, \cdots, \bar{\mathbf{r}}_{Nc}, \dot{\mathbf{R}}_1, \cdots, \dot{\mathbf{R}}_1, \bar{\mathbf{V}}_1, \cdots, \bar{\mathbf{V}}_N\right) \\
&= -\frac{1}{2}\sum_{j=1}^{N} \left\langle \left(\int_{\partial \bar{B}_j} \bar{\Phi}\bar{\mathbf{n}}_j\, \bar{v}, \int_{\partial \bar{B}_j} \bar{\Phi}(\bar{\mathbf{l}}_j \times \bar{\mathbf{n}}_j) \right), (\bar{\mathbf{V}}_j, \bar{\Omega}_j) \right\rangle \\
&\qquad + \frac{1}{2}\sum_{j=1}^{N} \left\langle (\mathbf{V}_j, \Omega_j), \mathbf{M_b} \cdot (\mathbf{V}_j, \Omega_j) \right\rangle .
\end{aligned} \tag{6.34}
$$

Note the relations,

$$
\bar{\mathbf{V}}_j = \dot{\bar{\mathbf{r}}}_{jc} = \mathbf{R}_j \cdot \mathbf{V}_j, \quad \dot{\mathbf{R}}_j = \mathbf{R}_j \cdot \hat{\Omega}_j, \quad \bar{\Omega}_j = \mathbf{R}_j \cdot \Omega_j \quad j = 1, \cdots, N \tag{6.35}
$$

and also

$$
\begin{aligned}
\bar{\mathbf{l}}_j &= \mathbf{R}_j \cdot \mathbf{l}_j, \\
\bar{\mathbf{n}}_j &= \mathbf{R}_j \cdot \mathbf{n}_j .
\end{aligned}
$$

Proceed, in standard fashion, to obtain the Euler–Lagrange equations by implementing the least action principle,

$$
\delta \int_{t_1}^{t_2} \mathscr{L}\left(\mathbf{R}_1, \cdots, \mathbf{R}_N, \bar{\mathbf{r}}_{1c}, \cdots, \bar{\mathbf{r}}_{Nc}, \dot{\mathbf{R}}_1, \cdots, \dot{\mathbf{R}}_1, \bar{\mathbf{V}}_1, \cdots, \bar{\mathbf{V}}_N\right) dt = 0, \tag{6.36}
$$

for variations taken over curves $\delta q(t) \equiv (\delta\mathbf{R}_1(t), \cdots, \delta\mathbf{R}_N(t), \delta\bar{\mathbf{r}}_{1c}(t), \cdots, \delta\bar{\mathbf{r}}_{Nc}(t)) \in T_q(\mathrm{SE}(3))^N$ satisfying the fixed end-points condition: $\delta\mathbf{R}_i(t_1) = \delta\mathbf{R}_l(t_2) = 0$ and $\delta\bar{\mathbf{r}}_{ic}(t_1) = \delta\bar{\mathbf{r}}_{ic}(t_2) = 0$, $i = 1, \cdots, N$. Consider the first sum of fluid terms in the Lagrangian paired with $\bar{\mathbf{V}}_j$. For notational clarity, ignore the outside sum over all bodies for now.

$$
\begin{aligned}
&-\frac{1}{2}\int_{t_1}^{t_2}\delta\left\langle\int_{\partial\bar{B}_j}\bar{\Phi}\bar{\mathbf{n}}_j\,\bar{\nu},\bar{\mathbf{V}}_j\right\rangle_{\mathbb{R}^3}dt\\
&=-\frac{1}{2}\int_{t_1}^{t_2}\left\langle\int_{\partial\bar{B}_j}\sum_{k=1}^{N}\left\langle\frac{\partial\bar{\Phi}}{\partial\mathbf{R}_k}\bar{\mathbf{n}}_j,\delta\mathbf{R}_k\right\rangle_{T\,SO(3)}\bar{\nu}\right.\\
&\qquad\left.+\int_{\partial\bar{B}_j}\sum_{k=1}^{N}\left\langle\frac{\partial\bar{\Phi}}{\partial\bar{\mathbf{r}}_{kc}}\bar{\mathbf{n}}_j,\delta\bar{\mathbf{r}}_{kc}\right\rangle_{\mathbb{R}^3}\bar{\nu},\bar{\mathbf{V}}_j\right\rangle_{\mathbb{R}^3}dt\\
&-\frac{1}{2}\int_{t_1}^{t_2}\left\langle\int_{\partial\bar{B}_j}\bar{\Phi}\delta\mathbf{R}_j\cdot\mathbf{n}_j\,\bar{\nu},\bar{\mathbf{V}}_j\right\rangle_{\mathbb{R}^3}dt-\frac{1}{2}\int_{t_1}^{t_2}\left\langle\int_{\partial\bar{B}_j}\bar{\Phi}\bar{\mathbf{n}}_j\,\nu,\delta\dot{\bar{\mathbf{r}}}_{jc}\right\rangle_{\mathbb{R}^3}dt.
\end{aligned}
$$

Similarly, the sum of fluid terms in the Lagrangian paired with $\bar{\Omega}_j$ give

$$
\begin{aligned}
&-\frac{1}{2}\int_{t_1}^{t_2}\delta\left\langle\int_{\partial\bar{B}_j}\bar{\Phi}(\bar{\mathbf{l}}_j\times\bar{\mathbf{n}}_j)\,\bar{\nu},\bar{\Omega}_j\right\rangle_{\mathbb{R}^3}dt\\
&=-\frac{1}{2}\int_{t_1}^{t_2}\left\langle\int_{\partial\bar{B}_j}\sum_{k=1}^{N}\left\langle\frac{\partial\bar{\Phi}}{\partial\mathbf{R}_k}(\bar{\mathbf{l}}_j\times\bar{\mathbf{n}}_j),\delta\mathbf{R}_k\right\rangle_{T\,SO(3)}\bar{\nu}\right.\\
&\qquad\left.+\int_{\partial\bar{B}_j}\sum_{k=1}^{N}\left\langle\frac{\partial\bar{\Phi}}{\partial\bar{\mathbf{r}}_{kc}}(\bar{\mathbf{l}}_j\times\bar{\mathbf{n}}_j),\delta\bar{\mathbf{r}}_{kc}\right\rangle_{\mathbb{R}^3}\bar{\nu},\bar{\Omega}_j\right\rangle_{\mathbb{R}^3}dt\\
&-\frac{1}{2}\int_{t_1}^{t_2}\left\langle\int_{\partial\bar{B}_j}\bar{\Phi}\delta\mathbf{R}_j\cdot(\mathbf{l}_j\times\mathbf{n}_j)\,\bar{\nu},\bar{\Omega}_j\right\rangle_{\mathbb{R}^3}dt\\
&\qquad-\frac{1}{2}\int_{t_1}^{t_2}\left\langle\int_{\partial\bar{B}_j}\bar{\Phi}(\bar{\mathbf{l}}_j\times\bar{\mathbf{n}}_j)\,\nu,\delta\bar{\Omega}_j\right\rangle_{\mathbb{R}^3}dt.
\end{aligned}
$$

Reintroducing the summation over all bodies and carrying out the usual procedures, one gets for the pairing with $\delta\bar{\mathbf{r}}_{jc}, j=1,\cdots,N$:

$$
\begin{aligned}
&\frac{1}{2}\int_{t_1}^{t_2}\left\langle\delta\bar{\mathbf{r}}_{jc},\frac{d}{dt}\int_{\partial\bar{B}_j}\bar{\Phi}\bar{\mathbf{n}}_j\,\nu-\sum_{k=1}^{N}\int_{\partial\bar{B}_k}\frac{\partial\bar{\Phi}}{\partial\bar{\mathbf{r}}_{jc}}\bar{\mathbf{n}}_k\cdot\bar{\mathbf{V}}_k\,\nu-\sum_{k=1}^{N}\int_{\partial\bar{B}_k}\frac{\partial\bar{\Phi}}{\partial\bar{\mathbf{r}}_{jc}}\bar{\mathbf{n}}_k\cdot(\bar{\Omega}_k\times\bar{\mathbf{l}}_k)\,\nu\right\rangle_{\mathbb{R}^3}dt\\
&\equiv\frac{1}{2}\int_{t_1}^{t_2}\left\langle\delta\bar{\mathbf{r}}_{jc},\frac{d}{dt}\int_{\partial\bar{B}_j}\bar{\Phi}\bar{\mathbf{n}}_j\,\nu-\sum_{k=1}^{N}\int_{\partial\bar{B}_k}\frac{\partial\bar{\Phi}}{\partial\bar{\mathbf{r}}_{jc}}\bar{\mathbf{n}}_k\cdot\bar{\mathbf{U}}_k\,\nu\right\rangle_{\mathbb{R}^3}dt.
\end{aligned}
$$

Note the change in the subscript from j to k for certain quantities. This is due to the summation over all bodies. The terms paired with $\delta\mathbf{R}_j, j=1,\cdots,N$, are as follows:

$$
\begin{aligned}
&\int_{t_1}^{t_2}\left\langle-\frac{1}{4}\sum_{k=1}^{N}\int_{\partial\bar{B}_k}\text{trace}\left(\delta\mathbf{R}_j^T\cdot\mathbf{R}_j\cdot\mathbf{R}_j^{-1}\cdot\frac{\partial\bar{\Phi}}{\partial\mathbf{R}_j}\right)\bar{\mathbf{n}}_k\,\bar{\nu},\bar{\mathbf{V}}_k\right\rangle_{\mathbb{R}^3}dt\\
&-\frac{1}{2}\int_{t_1}^{t_2}\left\langle\mathbf{R}_j^{-1}\cdot\delta\mathbf{R}_j\cdot\int_{\partial\bar{B}_j}\bar{\Phi}\mathbf{n}_j\,\bar{\nu},\mathbf{R}_j^{-1}\cdot\bar{\mathbf{V}}_j\right\rangle_{\mathbb{R}^3}dt
\end{aligned}
$$

$$\int_{t_1}^{t_2} \left\langle -\frac{1}{4}\sum_{k=1}^{N} \int_{\bar{\partial} B_k} \text{trace}\left(\delta \mathbf{R}_j^T \cdot \mathbf{R}_j \cdot \mathbf{R}_j^{-1} \cdot \frac{\partial \bar{\Phi}}{\partial \mathbf{R}_j}\right) (\bar{\mathbf{l}}_k \times \bar{\mathbf{n}}_k)\, \nu, \bar{\Omega}_k \right\rangle_{\mathbb{R}^3} dt$$
$$-\frac{1}{2}\int_{t_1}^{t_2} \left\langle \mathbf{R}_j^{-1} \cdot \delta \mathbf{R}_j \cdot \int_{\bar{\partial} B_j} \bar{\Phi}(\mathbf{l}_j \times \mathbf{n}_j)\, \bar{\nu}, \mathbf{R}_j^{-1} \cdot \bar{\Omega}_j \right\rangle_{\mathbb{R}^3} dt$$
$$-\frac{1}{2}\int_{t_1}^{t_2} \left\langle \int_{\bar{\partial} B_j} \bar{\Phi}(\bar{\mathbf{l}}_j \times \bar{\mathbf{n}}_j)\, \bar{\nu}, \delta \mathbf{R}_j \cdot \Omega_j) \right\rangle_{\mathbb{R}^3} dt$$
$$-\frac{1}{2}\int_{t_1}^{t_2} \left\langle \int_{\bar{\partial} B_j} \bar{\Phi}(\bar{\mathbf{l}}_j \times \bar{\mathbf{n}}_j)\, \nu, \mathbf{R}_j \cdot \delta\Omega_j) \right\rangle_{\mathbb{R}^3} dt$$

$$= \int_{t_1}^{t_2} -\frac{1}{4}\,\text{trace}\left(\delta \mathbf{R}_j^T \cdot \sum_{k=1}^{N} \int_{\bar{\partial} B_k} \frac{\partial \bar{\Phi}}{\partial \mathbf{R}_j} (\bar{\mathbf{V}}_k + \bar{\Omega}_k \times \bar{\mathbf{l}}_k) \cdot \bar{\mathbf{n}}_k\, \bar{\nu}\right) dt$$
$$-\frac{1}{4}\int_{t_1}^{t_2} \text{trace}\left(\delta \mathbf{R}_j^T \cdot \mathbf{R}_j \cdot \int_{\bar{\partial} B_j} \widehat{\bar{\Phi}\mathbf{n}_j\, \bar{\nu} \times \mathbf{V}_j}\right) dt$$
$$-\frac{1}{2}\int_{t_1}^{t_2} \left\langle \kappa_j \times \int_{\bar{\partial} B_j} \bar{\Phi}(\mathbf{l}_j \times \mathbf{n}_j)\, \bar{\nu}, \Omega_{jj} \right\rangle_{\mathbb{R}^3} dt$$
$$-\frac{1}{2}\int_{t_1}^{t_2} \left\langle \int_{\bar{\partial} B_j} \bar{\Phi}(\mathbf{l}_j \times \mathbf{n}_j)\, \bar{\nu}, \kappa_j \times \Omega_j) \right\rangle_{\mathbb{R}^3} dt$$
$$-\frac{1}{4}\int_{t_1}^{t_2} \text{trace}\left(\delta \hat{\Omega}_j^{\,T} \cdot \int_{\bar{\partial} B_j} \widehat{\bar{\Phi}(\mathbf{l}_j \times \mathbf{n}_j)\, \bar{\nu}}\right) dt$$

$$= \int_{t_1}^{t_2} -\frac{1}{4}\,\text{trace}\left(\delta \mathbf{R}_j^T \cdot \sum_{k=1}^{N} \int_{\bar{\partial} B_k} \frac{\partial \bar{\Phi}}{\partial \mathbf{R}_j} \bar{\mathbf{n}}_k \cdot \bar{\mathbf{U}}_k\, \bar{\nu}\right) dt$$
$$-\frac{1}{4}\int_{t_1}^{t_2} \text{trace}\left(\delta \mathbf{R}_j^T \cdot \mathbf{R}_j \cdot \int_{\bar{\partial} B_j} \widehat{\bar{\Phi}\mathbf{n}_j\, \bar{\nu} \times \mathbf{V}_j}\right) dt$$
$$-\frac{1}{4}\int_{t_1}^{t_2} \text{trace}\left(\left(\mathbf{R}_j^{-1} \cdot \delta \dot{\mathbf{R}}_j - \mathbf{R}_j^{-1} \cdot \delta \mathbf{R}_j \cdot \hat{\Omega}_j\right)^T \cdot \int_{\bar{\partial} B_j} \widehat{\bar{\Phi}(\mathbf{l}_j \times \mathbf{n}_j)\, \bar{\nu}}\right) dt.$$

The last term is re-arranged just as in Kirchhoff's problem (refer (1.27) and the equations that follow). Gathering all terms one gets

$$\frac{1}{4}\,\text{trace}\left(\delta \mathbf{R}_j^T \cdot \left(\frac{d}{dt}\left(\mathbf{R}_j \cdot \int_{\bar{\partial} B_j} \widehat{\bar{\Phi}(\mathbf{l}_j \times \mathbf{n}_j)\, \bar{\nu}}\right) - \mathbf{R}_j \cdot \hat{\Omega}_j \cdot \int_{\bar{\partial} B_j} \widehat{\bar{\Phi}(\mathbf{l}_j \times \mathbf{n}_j)\, \bar{\nu}} \right.\right.$$
$$+ \mathbf{R}_j \cdot \widehat{\Omega_j \times \int_{\bar{\partial} B_j} \bar{\Phi}(\mathbf{l}_j \times \mathbf{n}_j)\, \bar{\nu}} + \mathbf{R}_j \cdot \widehat{\mathbf{V}_j \times \int_{\bar{\partial} B_j} \bar{\Phi}\mathbf{n}_j\, \bar{\nu}}$$
$$\left.\left. - \sum_{k=1}^{N} \int_{\bar{\partial} B_k} \frac{\partial \bar{\Phi}}{\partial \mathbf{R}_j} \bar{\mathbf{n}}_k \cdot \bar{\mathbf{U}}_k\, \bar{\nu}\right)\right), \quad j = 1, \cdots, N.$$

The Euler–Lagrange equations, after including the body terms, are therefore obtained as

$$\frac{d}{dt}\left(m_j\bar{\mathbf{V}}_j - \int_{\partial \bar{B}_j} \bar{\Phi}\bar{\mathbf{n}}_j\,\bar{v}\right) = -\sum_{k=1}^{N}\int_{\partial \bar{B}_k} \frac{\partial \bar{\Phi}}{\partial \bar{\mathbf{r}}_{jc}}\bar{\mathbf{n}}_k \cdot \bar{\mathbf{U}}_k\,\bar{v},$$

$$\mathbf{R}_j \cdot \left[\frac{d}{dt}\left(\widehat{I_j\Omega_j - \int_{\partial \bar{B}_j} \bar{\Phi}(\mathbf{l}_j \times \mathbf{n}_j)\,\bar{v}}\right) + \Omega_j \times \left(\widehat{I_j\Omega_j - \int_{\partial \bar{B}_j} \bar{\Phi}(\mathbf{l}_j \times \mathbf{n}_j)\,\bar{v}}\right)\right.$$

$$\left. + \mathbf{V}_j \times \widehat{\int_{\partial \bar{B}_j} \bar{\Phi}\mathbf{n}_j\,\bar{v}}\right] = -\sum_{k=1}^{N}\int_{\partial \bar{B}_k} \frac{\partial \bar{\Phi}}{\partial \mathbf{R}_j}\bar{\mathbf{n}}_k \cdot \bar{\mathbf{U}}_k\,\bar{v}, \quad j = 1, \cdots, N.$$

Introduce the notation,

$$\bar{\mathbf{L}}_j := m_j\bar{\mathbf{V}}_j - \int_{\partial \bar{B}_j} \bar{\Phi}\bar{\mathbf{n}}_j\,\bar{v} \equiv \mathbf{R}_j \cdot \mathbf{L}_j, \quad \bar{\mathbf{A}}_j := I_j\bar{\Omega}_j - \int_{\partial \bar{B}_j} \bar{\Phi}(\bar{\mathbf{l}}_j \times \bar{\mathbf{n}}_j)\,\bar{v} \equiv \mathbf{R}_j \cdot \mathbf{A}_j.$$

To provide physical meaning to each $\bar{\mathbf{L}}_j$ and $\bar{\mathbf{A}}_j$, one must interpret the integrals $-\int_{\partial \bar{B}_j} \bar{\Phi}\bar{\mathbf{n}}_j\,\bar{v}$ and $-\int_{\partial \bar{B}_j} \bar{\Phi}(\mathbf{l}_j \times \mathbf{n}_j)\,\bar{v}$. Focusing on the former, it is easy to see applying the integral theorems of vector calculus that it is not exactly the same as the fluid linear momentum of the flow with each of the other bodies stationary, since there are corresponding boundary integral terms on the other bodies as well. However, if the other bodies are removed and $\bar{\Phi}$ extended harmonically to the interior of the body domains, then the boundary integral does represent the linear momentum of this mathematically extended flow.

The E-L equations can now be written in the more compact form as

$$\frac{d\bar{\mathbf{L}}_j}{dt} = -\sum_{k=1}^{N}\int_{\partial \bar{B}_k} \frac{\partial \bar{\Phi}}{\partial \bar{\mathbf{r}}_{jc}}\bar{\mathbf{n}}_k \cdot \bar{\mathbf{U}}_k\,\bar{v},$$

$$\mathbf{R}_j \cdot \left[\frac{d\hat{\mathbf{A}}_j}{dt} + \widehat{\Omega_j \times \mathbf{A}_j} + \widehat{\mathbf{V}_j \times \mathbf{L}_j}\right] = -\sum_{k=1}^{N}\int_{\partial \bar{B}_k} \frac{\partial \bar{\Phi}}{\partial \mathbf{R}_j}\bar{\mathbf{n}}_k \cdot \bar{\mathbf{U}}_k\,\bar{v},$$

$$j = 1, \cdots, N. \tag{6.37}$$

As in Kirchhoff's equations the left hand sides of the angular momentum equations can also be written as $d/d(\bar{\mathbf{A}}_j + \bar{\mathbf{r}}_{jc} \times \bar{\mathbf{L}}_j)$. The terms on the right of each equation may be viewed as force and moment terms arising from the dynamic interaction with the other bodies. The complete system is obtained by appending equations (6.33) and (6.35) to the above and evolving them simultaneously from given initial positions and velocities of the bodies. Note that computing $\partial \bar{\Phi}/\partial \bar{\mathbf{r}}_{jc}$ and $\partial \bar{\Phi}/\partial \partial \mathbf{R}_j$, in the force and moment terms, might entail additional computational effort.

The sums $\sum_{j=1}^{N}\bar{\mathbf{L}}_j$ and $\sum_{j=1}^{N}\bar{\mathbf{A}}_j+\bar{\mathbf{r}}_{jc}\times\bar{\mathbf{L}}_j$ obviously represent the total system momenta in the spatially-fixed frame. Summing over j in (6.37) leads to the double sums

$$\sum_{j=1}^{N}\sum_{k=1}^{N}\int_{\partial\bar{B}_k}\frac{\partial\bar{\Phi}}{\partial\bar{\mathbf{r}}_{jc}}\bar{\mathbf{n}}_k\cdot\bar{\mathbf{U}}_k\,\bar{\nu},\quad \sum_{j=1}^{N}\sum_{k=1}^{N}\int_{\partial\bar{B}_k}\frac{\partial\bar{\Phi}}{\partial\mathbf{R}_j}\bar{\mathbf{n}}_k\cdot\bar{\mathbf{U}}_k\,\bar{\nu}.$$

Using the property of $\bar{\Phi}$ mentioned previously, namely, that it depends only on the relative positions and orientations of the bodies, one can then show that there is a pairwise cancellation of all terms in these double sums leading to the expected conservation laws for the system momenta:

$$\frac{d}{dt}\left(\sum_{j=1}^{N}\bar{\mathbf{L}}_j\right)=0,$$

$$\frac{d}{dt}\left(\sum_{j=1}^{N}\bar{\mathbf{A}}_j+\bar{\mathbf{r}}_{jc}\times\bar{\mathbf{L}}_j\right)=0.$$

6.3 A Single Buoyant Rigid Body Above an Impermeable Flat Boundary

The above problem also allows one to consider the problem of a single rigid body above an impermeable flat bottom boundary, as shown in the left of Fig. 6.4. Recalling the discussion from Chap. 3 on image fields, it is obvious that the flow is equivalent to a two-rigid body problem in an unbounded domain in which the flat boundary is a line of reflection symmetry, as shown in the right of Fig. 6.4. It is easy to see that a two-body configuration that satisfies this symmetry condition at initial time continues to do so for all times. The impenetrable, Neumann boundary condition at the flat boundary is therefore satisfied at all times.

Labeling the real body as 1 and its image as 2, it is easy to see that the coordinates of the centers of masses obey

$$\bar{x}_{1c}(t)=\bar{x}_{2c}(t),\quad \bar{y}_{1c}(t)=\bar{y}_{2c}(t),\quad \bar{z}_{1c}(t)=-\bar{z}_{2c}(t)\quad\forall t$$

and the rotation matrices satisfy

$$\left(\mathbf{R}_{1\alpha}\cdot\bar{\mathbf{X}},\mathbf{R}_{1\beta}\cdot\bar{\mathbf{X}},\mathbf{R}_{1\gamma}\cdot\bar{\mathbf{X}}\right)=\left(\mathbf{R}_{2\alpha}\cdot\bar{\mathbf{X}},\mathbf{R}_{2\beta}\cdot\bar{\mathbf{X}},-\mathbf{R}_{2\gamma}\cdot\bar{\mathbf{X}}\right)\quad\forall t,$$

where $\bar{\mathbf{r}}_{1c}\equiv(\bar{x}_{1c},\bar{y}_{1c},\bar{z}_{1c})^T$, etc., and $\mathbf{R}_1\equiv\left(\mathbf{R}_{1\alpha},\mathbf{R}_{1\beta},\mathbf{R}_{1\gamma}\right)^T$, etc. To obtain the Euler–Lagrange equations, substitute $N=2$ in (6.34) and take variations consistent

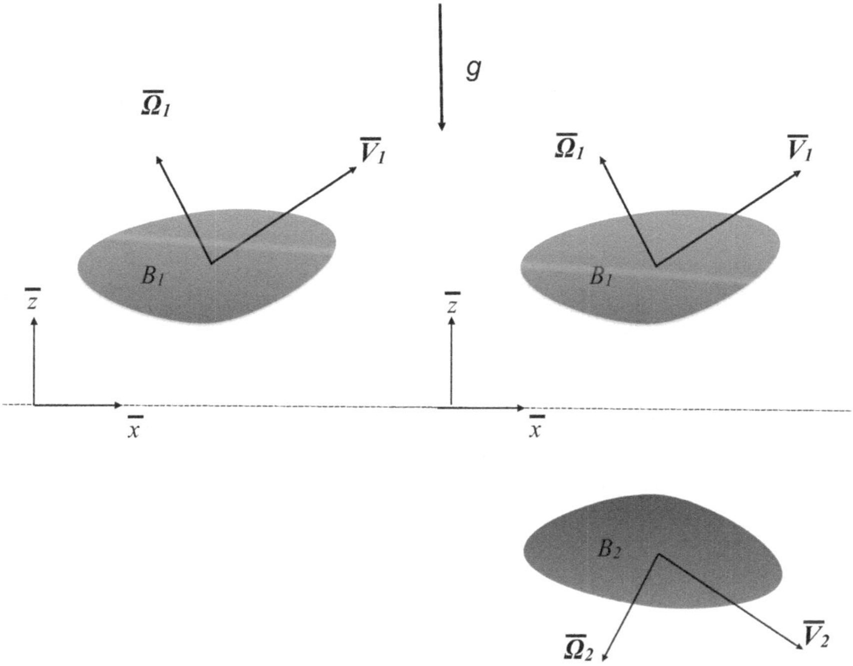

Fig. 6.4 The dynamically coupled problem of a single rigid body in the proximity of a flat rigid impenetrable surface (left) can be viewed as the dynamically coupled problem of two-rigid bodies in an infinite domain (right), assuming the bodies satisfy the reflection symmetry condition at time $t = 0$

with the above relations. For example, $\delta\bar{x}_{1c} = \delta\bar{x}_{2c}, \delta\bar{y}_{1c} = \delta\bar{y}_{2c}, \delta\bar{z}_{1c}(t) = -\delta\bar{z}_{2c}(t)$. The system (6.37) now reduces to

$$\begin{aligned} \frac{d\bar{\mathbf{L}}_1}{dt} &= -2\int_{\partial\bar{B}_1} \frac{\partial\bar{\Phi}}{\partial\bar{\mathbf{r}}_{1c}}\bar{\mathbf{n}}_1 \cdot \bar{\mathbf{U}}_1\, \bar{v}, \\ \mathbf{R}_1 \cdot \left[\frac{d\hat{\mathbf{A}}_1}{dt} + \widehat{\Omega_1 \times \mathbf{A}_1} + \widehat{\mathbf{V}_1 \times \mathbf{L}_1}\right] &= -2\int_{\partial\bar{B}_1} \frac{\partial\bar{\Phi}}{\partial\mathbf{R}_1}\bar{\mathbf{n}}_1 \cdot \bar{\mathbf{U}}_1\, \bar{v}, \end{aligned} \tag{6.38}$$

and the right sides now represent the external force and moment coming from the presence of the wall and, in fact, have non-zero components only in the y-direction, i.e. in the wall-perpendicular direction.

This configuration also allows one to include the effects of positive or negative buoyancy. One assumes in such a case that the extent of the fluid is infinite in the upward direction or is bounded by another solid or free surface boundary. In any case, the fluid at some large but finite distance upward from the body is assumed to be undisturbed. In the case of a free surface at the top, for example, referring to Sect. 6.1.1 this implies $\eta(t) = \eta_0$ for all t and that $\bar{\nabla}\bar{\Phi} \cdot \bar{\mathbf{n}} = 0$ there. The problem then becomes, in a sense, complementary to the problem considered in that section where it was assumed the bottom flat boundary is far enough as to have a negligible influence on the body+free surface dynamics. The corresponding integral terms in (6.11) are therefore set to zero. The Lagrangian is defined as usual as the kinetic energy minus the potential energy. Variations in the potential energy term $(\rho_b - \rho_f)g \int_{B_1} \bar{z}\,\bar{\mu}$, are easily obtained from $\delta \int_{B_1} \bar{z}\,\bar{\mu} = \delta \int_{B_1} \bar{\mathbf{r}}_{1c} \cdot \bar{\mathbf{k}}\,\bar{\mu} = \delta \bar{\mathbf{r}}_{1c} \cdot \bar{\mathbf{k}} \int_{B_1} \bar{\mu}$, leading to the equations

$$\frac{d\bar{\mathbf{L}}_1}{dt} = -2 \int_{\partial \bar{B}_1} \frac{\partial \bar{\Phi}}{\partial \bar{\mathbf{r}}_{1c}} \bar{\mathbf{n}}_1 \cdot \bar{\mathbf{U}}_1\,\bar{\nu} - (\rho_b - \rho_f) g \mathscr{V}_B \bar{\mathbf{k}},$$

$$\mathbf{R}_1 \cdot \left[\frac{d\hat{\mathbf{A}}_1}{dt} + \widehat{\Omega_1 \times \mathbf{A}_1} + \widehat{\mathbf{V}_1 \times \mathbf{L}_1} \right] = -2 \int_{\partial \bar{B}_1} \frac{\partial \bar{\Phi}}{\partial \mathbf{R}_1} \bar{\mathbf{n}}_1 \cdot \bar{\mathbf{U}}_1\,\bar{\nu},$$

where $\mathscr{V}_B$ denotes the volume of the body. Note that due to the assumption of uniform body and fluid densities, the force of gravity and the force of buoyancy both act through the body's center of mass and therefore cause no moment about any point. In the limit of the bottom boundary also moving to infinity, one recovers (1.29) with the added net buoyancy force term on the right.

Dynamically Coupled Free Surface Plus Vortices

Strictly speaking this problem is not an extension of Kirchhoff's model due to the absence of a rigid body. Nevertheless, as long as vortices are shed from rigid bodies in the vicinity of a free surface, such interactions need to be analyzed too. Moreover, there are also several examples of vortices intersecting the free surface [104, 141], ranging from vortices in the upper layers of planetary atmospheres to more familiar examples of bathtub and swimming pool vortices. In many cases, these vortices are shed off the free surface itself, just as in the case of solid surfaces. Such interactions are invariably in a dynamically coupled setting which makes the problem intrinsically relevant to the topic of this book.

The most comprehensive treatment of this problem in the framework of geometric mechanics is due to Lewis, Marsden, Montgomery, and Ratiu [96] who considered the case of smooth vorticity fields that do not intersect the free surface. They showed that the basic bracket is a sum Poisson bracket comprising of the vorticity bracket and Zakharov's bracket. The vorticity bracket is essentially the Lie–Poisson bracket for incompressible, inviscid flows. Referring to the discussion at the beginning of Sect. 1.4, this bracket is obtained by symmetry reduction of the cotangent bundle of $G = \mathrm{Diff}_{\mathrm{vol}}$, the Lie group of volume-preserving diffeomorphisms of the fluid domain, and resides on the dual of the Lie algebra of this group; see also [111]. Using deep ideas from theories of reduction of Hamiltonian systems with symmetry they also described other representations of this basic bracket. Rouhi and Wright [142] examined the 2D motion of point vortices interacting with a free surface as also the 3D problem of vortex rings interacting with a free surface, and presented a Poisson bracket structure for each. Sum Poisson brackets are not mentioned but pairs of brackets are presented: the point vortex bracket and the Zakharov bracket for the 2D problem, and a local version of the Marsden–Weinstein ring bracket and the Zakharov bracket for the 3D problem. Kuznetsov and Ruban [91] also discuss the Hamiltonian formulation and analyze some other aspects of these interactions. In a non-geometric-mechanics framework, an interesting paper is due to Wu [190] who proposed a theory for vortices intersecting a fluid–fluid interface in the presence of viscosity.

Combinations

The models described so far in the book can be further combined with each other to give rise to new models. This would be not be purely an exercise in theoretical mechanics since there are situations where such combinations are relevant. For example, a rigid body near the free surface of water typically interacts with both the water waves and the vortices it sheds or that are shed off the free surface. With reference to the applications described in Chap. 1, an autonomous underwater vehicle entering or exiting the depths would experience such interactions. Extending such a model to the case of multiple rigid bodies is a natural next step. The dynamics of multiple bodies, whether rigid or deformable, has applications to another problem of long-standing interest: the formation swimming or flying of engineered vehicles—aircraft, ships, etc., or nature's creatures—fish, birds, etc., where multiple bodies move in close proximity of each other and vortices shed from one body can influence the motion of its neighbors. It is of course necessary to act control actuation in the models developed in the dynamically coupled setting, as stated previously in the book, before applying them to these real-world examples.

However, there are non-trivial theoretical challenges in developing these combinations, some of which have been known for a long time since they are independent of the dynamically coupled setting. For example, C. C. Lin's formulation described in Sect. 3.1 applies to multiply-connected domains but does not provide a method for obtaining the Kirchhoff–Routh analytically, even for the relatively simple case where the stationary rigid boundaries are two circular cylinders. Equivalently, it does not provide an analytic method for extending Milne-Thomson's circle theorem to construct the image field of point vortices external to two (or more) circles. It is only relatively recently that progress has been made on this topic [36, 37] using complex variable techniques. To the best of the author's knowledge, the corresponding problem of the image field of a vortex ring in multiple spheres, which would presumably extend Weiss' theorem, remains an open question.

Before closing this chapter, yet another application example that has recently emerged in the area of "green energy" is worth mentioning: a floating offshore wind turbine (FOWT), where vortices, waves, and bodies all interact with each other in partially-dynamically coupled and controlled settings. Plenty of illustrations can be easily found on the internet; a schematic sketch is shown in Fig. 6.5 (top). Popular mostly in Europe so far, the development of offshore wind turbines is motivated by the fact that stronger and more constant winds prevail over the oceans and seas than on land. These favorable features of the wind increase as one goes further seaward. However, the greater water depth makes fixing the wind turbine base to the sea-bed less feasible and hence the concept of a floating offshore wind turbine. "Floating" of course does not mean freely floating, rather the base platform is connected to the sea-bed by mooring lines or tethering cables which provide partial constraining and stability of the structure through tension in the lines. In addition, ballast mechanisms are sometimes used for stabilizing the structure. The dynamics of the structure and the water waves is therefore only partially coupled. Nevertheless, the coupled dynamics is still nonlinear largely due to the inherent nonlinearities present in the wave dynamics.

FOWTs are tremendous engineering challenges, both in modeling and in implementation. The coupled nonlinear dynamics of the rigid structure, waves, and vortices is more complex than in autonomous underwater vehicles and, in addition, there are the added dynamical effects due to the wind flow and the tethering system for the base. For example, the effect of the wind and the blade-tip vortices on the dynamics of the water surface. Indeed, the wind and vortices driven two-phase air–water system by itself is an interesting dynamical system. The high Reynolds numbers involved also imply the importance of wind turbulence and the rigid body has a more complex geometry and a bigger structure, with several sub-components, than that of AUVs. The coupled nonlinear dynamics of FOWTs is also more complex when compared with their land-based counterparts or even fixed offshore wind turbines. The nonlinear hydrodynamic coupling between the structure and the water waves, with the partial tethering constraints, results in motion of the structure. This in turn influences the turbine's interaction with the wind.

Attempting to construct a nonlinear model for the entire FOWT would be a very ambitious task, but subsystems can be extracted to which the previously developed models are relevant. For example, one could consider the configuration of a floating vertical cylinder with a base platform attached to tethering cables, as in Fig. 6.5 (bottom). The effects of the wind may be ignored, as a first approximation, and the focus could be on the coupled dynamics of the cylinder and vorticity-free water waves, with the tethering cables providing constraint forces. Such a model could be built based on the ideas in Sect. 6.1.1. These models could then be extended to include the effects of possible water vortices shed from the circular cylinder geometry, the trailing air vortices from the blade tips and the structural loading from the rotor assembly. Most current modeling efforts on the topic rely on linearized and empirical approaches, and nonlinear models are still largely lacking. These approaches typically study the structural response of the wind turbine to the different loading—aero, hydro, and due to the tethering cables—modeled in various approximate ways. Indeed the configuration suggested above, when the cylinder is slender, is often studied in the offshore structures literature by using an approximate equation referred to as "Morrison's equation." Jonkman's paper [78] gives a good introduction to the topic and also to the modeling and code development efforts.

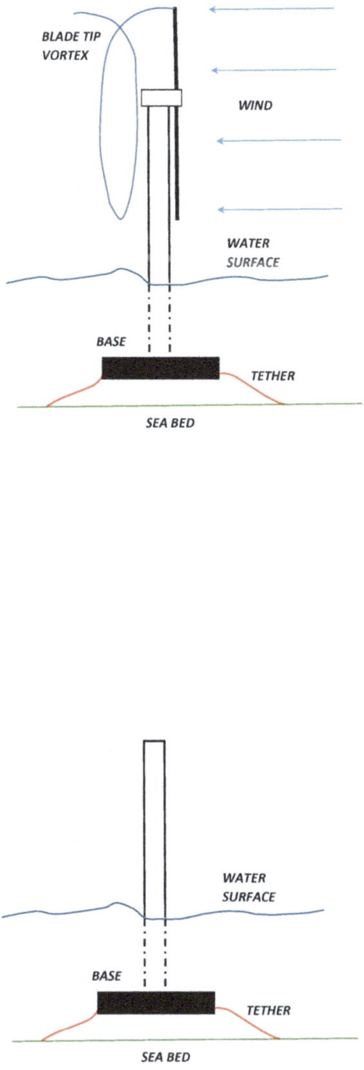

Fig. 6.5 Schematic sketches of an actual floating offshore wind turbine (top) and a simplified configuration (bottom)

Appendix A
A Brief Introduction to Geometric Mechanics

The subject of geometric mechanics builds upon and greatly expands classical mechanics theories using modern developments in differential geometry. This appendix provides a brief introduction to the subject for readers who are unfamiliar with it, and is primarily based upon the texts of [3, 105, 109] and [2]. The reader is also encouraged to consult [11, 12, 66, 159]. Some of these references also discuss the basic analysis and differential geometric concepts at length [3], such as differential forms and exterior algebra, but the reader may also look up texts on differential geometry like, for example, [38, 164, 180] for a better grasp of the basics. It goes without saying that almost every notion described below has associated with it a plethora of mathematical details which the interested reader may seek out in these cited texts or in others.

Theories in geometric mechanics generally have as a first step the identification and definition of a configuration space Q. In classical mechanics terminology, this is the space of "generalized coordinates." From a modern differential geometric viewpoint, Q has, at the very least, the structure of a manifold and in many applications comes equipped with more structure and features, such as an infinite degree of smoothness, a Riemannian metric, a symplectic form, Poisson brackets, etc. In this book it will be assumed that all manifolds are infinitely smooth. A manifold P generalizes the notion of vector spaces,[1] like $\mathbb{R}^N$, to spaces that are not linear but are locally like $\mathbb{R}^N$. A simple example being that of a curve. The set of points of the curve cannot be given the structure of a linear space but an infinitesimal portion of the curve can be identified with a line segment (as every student of calculus knows). Similarly for curved surfaces and hypersurfaces. But the utility of a manifold often lies in the fact that abstract spaces can be given the structure of a manifold, for example, the set of orthogonal 3×3 rotation matrices of a rigid body. The evolutions of a familiar mechanical system like a rigid body (pinned at a point) can then be identified with trajectories on this manifold. Computations and calculations involving manifolds are done using coordinate charts. For a given manifold coordinate

[1] Vector spaces are special cases of manifolds.

B. N. Shashikanth, *Dynamically Coupled Rigid Body-Fluid Flow Systems*,
https://doi.org/10.1007/978-3-030-82646-8

charts could be global, i.e. just one set of coordinates can be used for the entire manifold, or they could be local.

At any point $p \in P$, one can associate a linear space called the tangent space T_pP consisting of all vectors tangent to P and based at p. Invoking classical terminology again, T_qQ is the space of "generalized velocities" based at the point q. The set of all points in P together with all its tangent spaces is called the tangent bundle and denoted by TP. This is also a manifold and in fact a special kind of manifold called a fiber bundle. One thinks of the space T_pP as the fiber over the base point $p \in Q$, and the space TP as the union of the base space and all the fibers.

The dimension of a manifold can be infinite. Finite-dimensional manifolds are assumed in most of what follows since many of the basic ideas carry over formally to infinite-dimensional manifolds. Note that a finite-dimensional manifold can generally be embedded in a linear space of higher dimension, but the definition of a manifold itself does not require an ambient space. For an N-dimensional manifold P, T_pP is also N-dimensional and therefore isomorphic to $\mathbb{R}^N$. Invoking linear space theory, one can associate with each T_pP a dual space of co-vectors or 1-forms T_p^*P by defining a non-degenerate pairing between the two spaces. This results in another fiber bundle, the cotangent bundle T^*P. The non-degenerate pairing is a bilinear map $\langle , \rangle : T_pP \times T_p^*P \to \mathbb{R}$. Standard results in linear space theory guarantee that such a map can be constructed from the canonical inner product on $\mathbb{R}^N$. Charts for tangent and cotangent bundles are usually in pairs, one for the base space and one for the fiber.

The non-degenerate pairing allows one to define functional derivatives on a manifold P. For a function $F : P \to \mathbb{R}$, the functional derivative $\partial F/\partial p$ is defined by the relation

$$\lim_{\varepsilon \to 0} [F(p+\varepsilon\delta p) - F(p)] = \left\langle \frac{\partial F}{\partial p}, \varepsilon\delta p \right\rangle,$$

where δp is a variation based at the point p and is identified with an element of T_pP (the δ symbol is used in the functional derivative, instead of ∂, if P is infinite-dimensional). Recalling calculus in $\mathbb{R}^N$, this is essentially the generalization of the familiar relation between a directional derivative and the gradient of a function. It follows from this definition that the functional derivative is identified with an element of the dual space T_p^*P. For functions defined on tangent and cotangent bundles, such as Lagrangian and Hamiltonian functions, the above definition generalizes in obvious ways: for example, if $F : TP \to \mathbb{R}$, the functional derivative $\partial F/\partial p$ is defined by the relation

$$\lim_{\varepsilon \to 0} [F(p+\varepsilon\delta p, v_p) - F(p, v_p)] = \left\langle \frac{\partial F}{\partial p}, \varepsilon\delta p \right\rangle,$$

and $\partial F/\partial v_p$ is defined by the relation

$$\lim_{\varepsilon \to 0} [F(p, v_p+\varepsilon\delta v_p) - F(p, v_p)] = \left\langle \frac{\partial F}{\partial v_p}, \varepsilon\delta v_p \right\rangle.$$

In the above relation, $\delta v_p \in T_{v_p}T_pP$; however, since T_pP is already a vector space its tangent space is an identical copy, and one again makes the identification $\delta v_p \in T_pP$ and $\partial F/\partial v_p \in T_p^*P$.

Another important notion is that of a vector field X on a manifold. This is the assignment of a vector at each point of the manifold, $X : P \to T_pP$.

Symplectic and Poisson Manifolds

A symplectic manifold is a manifold P equipped with a symplectic form, i.e. a closed, non-degenerate 2-form $\Omega : TP \times TP \to \mathbb{R}$. Recall that a 2-form on a vector space V is a bilinear, skew-symmetric mapping $\Lambda : V \times V \to \mathbb{R}$. A 2-form on a manifold defines such a map on two copies of the tangent space at each point of P, $\Lambda(p) : T_pP \times T_pP \to \mathbb{R}$.

A Poisson manifold is a manifold equipped with a Poisson bracket which is defined on $C^\infty(P,R)$ the space of smooth real-valued functions on P. A Poisson bracket $\{,\} : C^\infty(P,\mathbb{R}) \times C^\infty(P,\mathbb{R}) \to \mathbb{R}$ satisfies four properties: bilinearity, skew-symmetry, Leibniz's rule, and Jacobi's identity. Poisson and symplectic manifolds bear an intimate relation to each other. Any symplectic manifold is also a Poisson manifold, with the symplectic form and Poisson bracket related as:

$$\Omega(X_F, X_G)(p) = \{F,G\}(p).$$

More generally, a Poisson manifold consists of symplectic submanifolds termed symplectic leaves.

Cotangent bundles are examples of both symplectic and Poisson manifolds since they come equipped with a canonical symplectic form and a canonical Poisson bracket. This means that for a $2N$-dimensional cotangent bundle T^*Q there exists local charts with coordinates $(q_1, \cdots, q_N, \alpha_1, \cdots, \alpha_N)$ representing points $(q, \alpha_q) \in T^*Q$ in which the symplectic form and Poisson bracket can be written as

$$\Omega(q, \alpha_q) = \sum_{k=1}^{N} dq_k \wedge d\alpha_k, \quad \{F,G\} = \sum_{k=1}^{N} \left\{ \frac{\partial F}{\partial q_k}\frac{\partial G}{\partial \alpha_k} - \frac{\partial F}{\partial \alpha_k}\frac{\partial G}{\partial q_k} \right\}.$$

Lagrangian and Hamiltonian Systems

The famous least action principle in mechanics states that a system evolving on a configuration manifold Q, with trajectory given by the curve $q(t), t \in [t_i, t_f]$ between fixed points $q(t_i) = q_i, q(t_f) = q_f \in Q$, satisfies

$$\delta \int_{t_i}^{t_f} L(q, \dot{q})\, dt = 0,$$

for all variations $\delta(q(t)) \in T_{q(t)}Q$ of the curve satisfying the fixed end-points condition: $\delta q_i = \delta q_f = 0$. The function $L : TQ \times \mathbb{R} \to \mathbb{R}$ is called the Lagrangian of the system. Implementing the least action principle leads to the Euler–Lagrange equations:

$$\frac{d}{dt}\left(\frac{\partial L}{\partial \dot{q}}\right) - \frac{\partial L}{\partial q} = 0.$$

If Q is N-dimensional with generalized coordinates $(q_1, \cdots, q_N)$, then one obtains N equations for the Euler–Lagrange system:

$$\frac{d}{dt}\left(\frac{\partial L}{\partial \dot{q}_k}\right) - \frac{\partial L}{\partial q_k} = 0, \quad k = 1, \cdots, N.$$

Referring to the discussion on functional derivatives, $\partial L/\partial \dot{q} \in T_q^*Q$. The Lagrangian is said to be degenerate if, essentially, the mapping $\dot{q} \to \partial L/\partial \dot{q}$ is not invertible. Define a new set of system variables called the canonical momenta[2] $p_q \in T_q^*Q$, as $p_q := \partial L/\partial \dot{q}$. For a non-degenerate Lagrangian this leads to a set of $2N$ ODE on the cotangent bundle T^*Q called Hamilton's equations:

$$\frac{dq_k}{dt} = \frac{\partial H}{\partial p_k}, \quad \frac{dp_k}{dt} = -\frac{\partial H}{\partial q_k}, \quad k = 1, \cdots, N,$$

where $H : T^*Q \to \mathbb{R}$ is called the Hamiltonian function of the system and is related to the Lagrangian by the Legendre transform:

$$H(p_q, q) := \langle p_q, \dot{q} \rangle_Q - L(q, \dot{q}),$$

where $\langle p_q, \dot{q} \rangle_Q$ denotes the non-degenerate pairing defined on Q. An important feature of Hamilton's equations is that the function H is conserved along the trajectories of the system on P, which can be easily verified from the form of the equations. The conservation of H implies that there exist submanifolds of P, given by the level surfaces of H, on which the trajectories lie. There are also non-autonomous Hamiltonian systems in which H has explicit time dependency and in such systems H may not be conserved along trajectories. Such systems are not considered in this book.

In the modern framework, Hamiltonian mechanics is viewed as occurring on symplectic and Poisson manifolds. This generalization allows a larger number of systems, including infinite-dimensional ones, to be identified as Hamiltonian systems without restricting to the canonical form of ODE above. The Hamiltonian vector field $X_H : P \to TP$ of a function $H : P \to \mathbb{R}$ on the symplectic manifold P satisfies the relation:

$$\Omega(X_H, v_p)(p) = \mathbf{d}H(p)(v_p), \quad p \in P, \; v_p \in T_pP,$$

[2] The traditional notation for the canonical momenta is used here instead of the symbol α_q used elsewhere in this appendix.

where $\mathbf{d}H : TP \to \mathbb{R}$ is the 1-form obtained from the function H. The Hamiltonian vector field $X_H : P \to TP$ of a function $H : P \to \mathbb{R}$ on the Poisson manifold P satisfies the relation:

$$\frac{dF}{dt}(p) = \{F,H\}(p), \quad p \in P, \quad F : P \to \mathbb{R},$$

where dF/dt represents the rate of change of the function along the Hamiltonian vector field X_H.

Lie Groups, Symmetries, Conserved Quantities, and Reduction

A Lie group G is a manifold which also has a group structure, i.e. its elements satisfy group composition $\mathfrak{C} : G \times G \to G$. Recall that group composition must have, at the very least, the following properties: (i) associativity, (ii) an identity element, and (iii) the inverse element. The tangent space to the group at the identity element e is called the Lie algebra of the group and denoted by $\mathfrak{g}$. The dual space of $\mathfrak{g}$ is denoted by $\mathfrak{g}^*$ and referred to as the dual of the Lie algebra. This vector space is a particularly important Poisson manifold, since it is equipped with the Lie–Poisson bracket. Its symplectic leaves are termed coadjoint orbits.

The action of a Lie group G on a manifold is a map $A : G \times P \to P$, i.e. each $g \in G$ is associated with a map of P to P. The map associated with the identity element leaves every point in P invariant. However, for any $p \in P$ there could be more than one element of G that leaves p invariant. In fact these elements of G form a subgroup $G_p \subset G$ called the isotropy subgroup of p. Note that a Lie group can also act on itself. The action of a Lie group naturally leads to a fiber bundle structure for P. For any point $p \in P$ one can consider the set of points $\{A(g \cdot p) \mid g \in G\}$. Identifying each such set with a point, one obtains the base space of this fiber bundle denoted by P/G. A fiber bundle is identified with its projection map $\pi : P \to P/G$. Two particularly important canonically defined actions of G are the adjoint action $\mathrm{Ad} : G \times \mathfrak{g} \to \mathfrak{g}$ and the coadjoint action $\mathrm{Ad}^* : G \times \mathfrak{g}^* \to \mathfrak{g}^*$.

If a Lie group acting on a symplectic or Poisson manifold leaves a function $H : P \to \mathbb{R}$ invariant, i.e. $H(A \cdot p) = H(p), \forall p \in P$, then the corresponding Hamiltonian system is said to be symmetric with respect to the action of the group G. A famous result due to Emily Noether states that a Hamiltonian system with symmetry possesses additional conserved quantities; familiar examples being the conserved spatial momenta of a free rigid body. In the modern setting, these notions are generalized to notion of the momentum map $\mathbf{J} : P \to \mathfrak{g}^*$ and its conservation. Dynamical trajectories of Hamiltonian systems with symmetries are therefore constrained to lie on the level surfaces of H and $\mathbf{J}$. Additional conserved quantities called Casimirs, $\mathscr{C} : P \to \mathbb{R}$, may be associated with a Poisson manifold P. These are constant on symplectic leaves and Poisson commute with every $F \in C^\infty(P,\mathbb{R})$, i.e. $\{\mathscr{C},F\} = 0$.

The fiber bundle construction mentioned above is a purely geometric construction. The more interesting question is whether one can obtain reduced Hamiltonian systems on the symmetry reduced spaces P/G. Marsden and Weinstein [110] answered this question for symplectic manifolds P in the affirmative, incorporating the conservation of the momentum map in their reduction process. In particular, they showed that the base space of the fiber bundle $\pi_\mu : \mathbf{J}^{-1}(\mu) \to \mathbf{J}^{-1}(\mu)/G_\mu$, where $\mu \in \mathfrak{g}^*$ and $\mathbf{J}^{-1}(\mu)$ is the corresponding level set of the momentum map, inherits a symplectic structure from P with an associated reduced Hamiltonian system. $G_\mu \subset G$ is the isotropy subgroup with respect to the coadjoint action. This basic picture of symplectic reduction has been considerably expanded in recent years; details may be found in the monograph [106]. For reduction of Poisson manifolds, reduction in the Lagrangian framework and interesting dynamical features due to symmetry and reduction, see [32, 108, 109] and [107].

Appendix B
Leading Order Behavior of Velocity and Vector Potential Fields of a Curved Vortex Filament

A detailed evaluation of the leading order behavior of integrals (4.3) and (4.4), as a field point approaches the jth ring, is presented in this Appendix. The reader is also referred to [15, 29, 83, 147, 166].

To obtain this, refer to the toroidal neighborhood construction described in Sect. 4.1 and consider a field point p lying on one of the circles of radius ε (and therefore on the surface of the solid torus). Let $p_{\rm f}$ denote the center of this circle which lies on the jth ring. Denote by $s_{\rm f}$ the arc-length parameter for $p_{\rm f}$. Assign rectangular coordinates (x,y) and polar coordinates (r,θ) to the plane which contains this circle and, by extension, take as global rectangular coordinates in $\mathbb{R}^3$, (x,y,z), where z is along the unit tangent to the curve at $p_{\rm f}$. Use the Taylor's series representations:

$$
\begin{aligned}
\mathbf{r}(s_j) &= \mathbf{r}(s_{\rm f}) + (s_j - s_{\rm f})\frac{d\mathbf{r}}{ds_j}_{|s_{\rm f}} + \ldots, \\
\mathbf{t}_j(s_j) &= \mathbf{t}_j(s_{\rm f}) + (s_j - s_{\rm f})\frac{d\mathbf{t}_j}{ds_j}_{|s_{\rm f}} + \ldots,
\end{aligned}
$$

and denote by X_{sf}, Y_{sf}, Z_{sf} the components of $\mathbf{r}(s_j) - \mathbf{r}(s_{\rm f})$ in the global coordinate system. Set

$$
x(p) - x(s_{\rm f}) = \varepsilon\cos\theta, \quad y(p) - y(s_{\rm f}) = \varepsilon\sin\theta,
$$

B. N. Shashikanth, *Dynamically Coupled Rigid Body-Fluid Flow Systems*,
https://doi.org/10.1007/978-3-030-82646-8

and expand the integrand of (4.4) as

$$\frac{\mathbf{t}(s_j)\times(\mathbf{r}(p)-\mathbf{r}(s_j))}{|\mathbf{r}(p)-\mathbf{r}(s_j)|^3}$$

$$=\frac{\left(\mathbf{t}_j(s_f)+(s_j-s_f)\frac{d\mathbf{t}_j}{ds_j}_{|s_f}+\frac{(s_j-s_f)^2}{2}\frac{d^2\mathbf{t}_j}{ds_j^2}_{|s_f}+\ldots\right)\times\left(\varepsilon\cos\theta\mathbf{i}+\varepsilon\sin\theta\mathbf{j}-(s_j-s_f)\frac{d\mathbf{r}}{ds_j}_{|s_f}-\frac{(s_j-s_f)^2}{2}\frac{d^2\mathbf{r}}{ds_j^2}_{|s_f}-\ldots\right)}{\left(\varepsilon^2-2\varepsilon\cos\theta X_{sf}-2\varepsilon\sin\theta Y_{sf}+X_{sf}^2+Y_{sf}^2+Z_{sf}^2\right)^{3/2}},$$

$$=\frac{\left(\mathbf{t}_j(s_f)+(s_j-s_f)\frac{d\mathbf{t}_j}{ds_j}_{|s_f}+\frac{(s_j-s_f)^2}{2}\frac{d^2\mathbf{t}_j}{ds_j^2}_{|s_f}+\ldots\right)\times\left(\varepsilon\cos\theta\mathbf{i}+\varepsilon\sin\theta\mathbf{j}-(s_j-s_f)\frac{d\mathbf{r}}{ds_j}_{|s_f}-\frac{(s_j-s_f)^2}{2}\frac{d^2\mathbf{r}}{ds_j^2}_{|s_f}-\ldots\right)}{\left(\varepsilon^2-2\varepsilon\cos\theta\left(\frac{(s_j-s_f)^2}{2}\frac{d^2x}{ds_j^2}_{|s_f}+\ldots\right)-2\varepsilon\sin\theta\left(\frac{(s_j-s_f)^2}{2}\frac{d^2y}{ds_j^2}_{|s_f}+\ldots\right)+(s_j-s_f)^2+(s_j-s_f)^3\frac{d\mathbf{r}}{ds_j}_{|s_f}\cdot\frac{d^2\mathbf{r}}{ds_j^2}_{|s_f}+\ldots\right)^{3/2}}$$

In the denominator of the last equation simplifications occur since for arc-length parametrization, $d\mathbf{r}/ds_j=\mathbf{t}_j$, and from the construction of the coordinates, $\mathbf{i}\cdot d\mathbf{r}/ds_{j|s_f}=\mathbf{j}\cdot d\mathbf{r}/ds_{j|s_f}=0$. Moreover, $(dx/ds_j)^2+(dy/ds_j)^2+(dz/ds_j)^2=<\mathbf{t}_j,\cdot\mathbf{t}_j>=1$.

Define

$$k:=\frac{s_j-s_f}{\varepsilon},\tag{B.1}$$

and let $\varepsilon\to 0$. This gives

$$\frac{\mathbf{t}(s_j)\times(\mathbf{r}(p)-\mathbf{r}(s_j))}{|\mathbf{r}(p)-\mathbf{r}(s_j)|^3}$$

$$=\frac{1}{\varepsilon^3(1+k^2)^{3/2}}\frac{\left(\mathbf{t}_j(s_f)+\varepsilon k\frac{d\mathbf{t}_j}{ds_j}_{|s_f}+\frac{\varepsilon^2k^2}{2}\frac{d^2\mathbf{t}_j}{ds_j^2}_{|s_f}+\ldots\right)\times\left(\varepsilon\cos\theta\mathbf{i}+\varepsilon\sin\theta\mathbf{j}-\varepsilon k\mathbf{t}_j(s_f)-\varepsilon^2k^2\frac{d\mathbf{t}_j}{ds_j}_{|s_f}-\ldots\right)}{\left(1-\frac{2}{\varepsilon(1+k^2)}\cos\theta\left(\frac{\varepsilon^2k^2}{2}\frac{d^2x}{ds_j^2}_{|s_f}+\ldots\right)-\frac{2}{\varepsilon(1+k^2)}\sin\theta\left(\frac{\varepsilon^2k^2}{2}\frac{d^2y}{ds_j^2}_{|s_f}+\ldots\right)+\frac{\varepsilon^3k^3\mathbf{t}_j(s_f)\cdot\frac{d\mathbf{t}_j}{ds_j}_{|s_f}}{\varepsilon^2(1+k^2)}+\ldots\right)^{3/2}}$$

$$=\frac{1}{\varepsilon^3(1+k^2)^{3/2}}\left[\left(\mathbf{t}_j(s_f)+\varepsilon k\frac{d\mathbf{t}_j}{ds_j}_{|s_f}+\frac{\varepsilon^2k^2}{2}\frac{d^2\mathbf{t}_j}{ds_j^2}_{|s_f}+\ldots\right)\times\left(\varepsilon\cos\theta\mathbf{i}+\varepsilon\sin\theta\mathbf{j}-\varepsilon k\mathbf{t}_j(s_f)-\frac{\varepsilon^2k^2}{2}\frac{d\mathbf{t}_j}{ds_j}_{|s_f}-O(\varepsilon^3)\right)\right.$$

$$\left.\cdot\left(1+\frac{3}{\varepsilon(1+k^2)}\cos\theta\left(\frac{\varepsilon^2k^2}{2}\frac{d^2x}{ds_j^2}_{|s_f}+O(\varepsilon^3)\right)+\frac{3}{\varepsilon(1+k^2)}\sin\theta\left(\frac{\varepsilon^2k^2}{2}\frac{d^2y}{ds_j^2}_{|s_f}+O(\varepsilon^3)\right)-\frac{3}{2}\frac{\varepsilon^3k^3\mathbf{t}_j(s_f)\cdot\frac{d\mathbf{t}_j}{ds_j}_{|s_f}}{\varepsilon^2(1+k^2)}+O(\varepsilon^2)\right)\right]$$

$$=\frac{1}{\varepsilon^3(1+k^2)^{3/2}}\left[\left(\varepsilon\cos\theta\mathbf{j}-\varepsilon\sin\theta\mathbf{i}+\varepsilon^2k\frac{d\mathbf{t}_j}{ds_j}_{|s_f}\times(\cos\theta\mathbf{i}+\sin\theta\mathbf{j})+\frac{\varepsilon^2k^2}{2}\left(\mathbf{t}_j(s_f)\times\frac{d\mathbf{t}_j}{ds_j}_{|s_f}\right)+O(\varepsilon^3)\right)\right.$$

$$\left.\cdot\left(1+\left(\frac{3\varepsilon k^2}{2(1+k^2)}\cos\theta\frac{d^2x}{ds_j^2}_{|s_f}+O(\varepsilon^2)\right)+\left(\frac{3\varepsilon k^2}{2(1+k^2)}\sin\theta\frac{d^2y}{ds_j^2}_{|s_f}+O(\varepsilon^2)\right)-\frac{3\varepsilon}{2}\frac{\mathbf{t}_j(s_f)\cdot\frac{d\mathbf{t}_j}{ds_j}_{|s_f}}{1+k^2}+O(\varepsilon^2)\right)\right].\tag{B.2}$$

Before proceeding further, note from the Serret–Frenet equations that at any point of the curve

$$\left(\frac{d^2\mathbf{r}}{ds_j^2}\equiv\right)\frac{d\mathbf{t}_j}{ds_j}=\kappa\mathbf{n}_j,$$

where $\mathbf{n}_j$ is the principal unit normal of the curve and κ is the principal curvature. It follows that in (B.2), the last term is equal to zero.

Next, split $\mathbf{v}_j$ into the sum of a local contribution that comes from a portion of arc-length, say l, of the ring on either side of p_{f}, and an $O(1)$ contribution from the rest of the ring,

$$\mathbf{v}_j = \mathbf{v}_{j,local} + \mathbf{v}_{j,non-local},$$

so that

$$\begin{aligned}\mathbf{v}_{j,local}(p) &= \frac{\Gamma_j}{4\pi}\int_{s_{\mathrm{f}}-l}^{s_{\mathrm{f}}+l} \frac{\mathbf{t}(s_j)\times(\mathbf{r}(p)-\mathbf{r}(s_j))}{\left|\mathbf{r}(p)-\mathbf{r}(s_j)\right|^3} ds_j,\\ &= \frac{\Gamma_j}{4\pi}\int_{-l/\varepsilon}^{l/\varepsilon} \frac{\mathbf{t}(s_j)\times(\mathbf{r}(p)-\mathbf{r}(s_j))}{\left|\mathbf{r}(p)-\mathbf{r}(s_j)\right|^3} \varepsilon dk. \qquad \text{(B.3)}\end{aligned}$$

Within the rectangular brackets in (B.2) there are $O(\varepsilon)$ and $O(\varepsilon^2)$ terms. When the series is substituted in the integral above, the $O(\varepsilon)$ term gives rise to the following integral in k:

$$\int_{-l/\varepsilon}^{l/\varepsilon} \frac{1}{(1+k^2)^{3/2}} dk,$$

and the $O(\varepsilon^2)$ term gives rise to the following three integrals in k:

$$\int_{-l/\varepsilon}^{l/\varepsilon} \frac{k}{(1+k^2)^{3/2}} dk, \quad \int_{-l/\varepsilon}^{l/\varepsilon} \frac{k^2}{(1+k^2)^{3/2}} dk, \quad \int_{-l/\varepsilon}^{l/\varepsilon} \frac{k^2}{(1+k^2)^{5/2}} dk.$$

Resolve these integrals to obtain

$$\begin{aligned}\int_{-l/\varepsilon}^{l/\varepsilon} \frac{1}{(1+k^2)^{3/2}} dk &= \frac{k}{\sqrt{1+k^2}} \Big|_{-l/\varepsilon}^{l/\varepsilon} = \frac{2l}{\sqrt{\varepsilon^2+l^2}},\\ \int_{-l/\varepsilon}^{l/\varepsilon} \frac{k}{(1+k^2)^{3/2}} dk &= -\frac{1}{\sqrt{1+k^2}} \Big|_{-l/\varepsilon}^{l/\varepsilon} = 0,\\ \int_{-l/\varepsilon}^{l/\varepsilon} \frac{k^2}{(1+k^2)^{3/2}} dk &= \left(-\frac{k}{\sqrt{1+k^2}} + \sinh^{-1}(k)\right) \Big|_{-l/\varepsilon}^{l/\varepsilon},\\ &= -\frac{2l}{\sqrt{\varepsilon^2+l^2}} + \log\left(k+\sqrt{k^2+1}\right) \Big|_{-l/\varepsilon}^{l/\varepsilon},\\ &= -\frac{2l}{\sqrt{\varepsilon^2+l^2}} + \log\left(l+\sqrt{l^2+\varepsilon^2}\right) - \log\left(-l+\sqrt{l^2+\varepsilon^2}\right),\\ \int_{-l/\varepsilon}^{l/\varepsilon} \frac{k^2}{(1+k^2)^{5/2}} dk &= \frac{k^3}{3(1+k^2)^{3/2}} \Big|_{-l/\varepsilon}^{l/\varepsilon}\\ &= \frac{2l^3}{3(\varepsilon^2+l^2)^{3/2}}.\end{aligned}$$

Using the above results, one obtains the leading order terms in (B.3), after some analysis, as

$$\mathbf{v}_{j,local}(p) = \frac{\Gamma_j}{4\pi}\left[\frac{2l}{\sqrt{\varepsilon^2+l^2}}\left(\frac{\cos\theta}{\varepsilon}\mathbf{j}-\frac{\sin\theta}{\varepsilon}\mathbf{i}\right) + \left(-\frac{2l}{\sqrt{\varepsilon^2+l^2}}+\log\left(l+\sqrt{l^2+\varepsilon^2}\right)-\log\left(-l+\sqrt{l^2+\varepsilon^2}\right)\right)\frac{\kappa\mathbf{b}_j}{2}+O(1)\right],$$

where $\mathbf{b}_j = \mathbf{t}_j \times \mathbf{n}_j$ is the unit vector field in the binormal directions of the curve. These two vector fields can be identified with a point vortex field in the plane of the circle and a binormal vector field, and are $O(1/\varepsilon)$ and $O(\log\varepsilon)$, respectively. Since, $\mathbf{v}_{j,non-local}$ is $O(1)$ these two vector fields also characterize the leading order behavior of $\mathbf{v}_j$, as $\varepsilon \to 0$.

Evaluate the leading order behavior of (4.3) in a similar manner:

$$\frac{\mathbf{t}(s_j)}{|\mathbf{r}(p)-\mathbf{r}(s_j)|}$$

$$= \frac{1}{\varepsilon(1+k^2)^{1/2}}\left[\left(\mathbf{t}_j(s_f)+\varepsilon k\frac{d\mathbf{t}_j}{ds_j}_{|s_f}+\frac{\varepsilon^2k^2}{2}\frac{d^2\mathbf{t}_j}{ds_j^2}_{|s_f}+\dots\right)\right.$$

$$\left.\cdot\left(1+\frac{\cos\theta}{\varepsilon(1+k^2)}\left(\frac{\varepsilon^2k^2}{2}\frac{d^2x}{ds_j^2}_{|s_f}+O(\varepsilon^3)\right)+\frac{\sin\theta}{\varepsilon(1+k^2)}\left(\frac{\varepsilon^2k^2}{2}\frac{d^2y}{ds_j^2}_{|s_f}+O(\varepsilon^3)\right)-\frac{1}{2}\frac{\varepsilon^3k^3\mathbf{t}_j(s_f)\cdot\frac{d\mathbf{t}_j}{ds_j}_{|s_f}}{\varepsilon^2(1+k^2)}+O(\varepsilon^2)\right)\right]$$

$$= \frac{1}{\varepsilon(1+k^2)^{1/2}}\left[\left(\mathbf{t}_j(s_f)+\varepsilon k\frac{d\mathbf{t}_j}{ds_j}_{|s_f}+\frac{\cos\theta}{1+k^2}\frac{\varepsilon k^2}{2}\frac{d^2x}{ds_j^2}_{|s_f}\mathbf{t}_j(s_f)+\frac{\sin\theta}{1+k^2}\frac{\varepsilon k^2}{2}\frac{d^2y}{ds_j^2}_{|s_f}\mathbf{t}_j(s_f)+O(\varepsilon^2)\right)\right].$$

Split $\mathbf{A}_j$, like $\mathbf{v}_j$, into a local and non-local part,

$$\mathbf{A}_j = \mathbf{A}_{j,local} + \mathbf{A}_{j,non-local},$$

with

$$\mathbf{A}_{j,local}(p) = \frac{\Gamma_j}{4\pi}\int_{-l/\varepsilon}^{l/\varepsilon}\frac{\mathbf{t}_j(s_j)}{|\mathbf{r}(p)-\mathbf{r}(s_j)|}\varepsilon dk,$$

Noting that

$$\int_{-l/\varepsilon}^{l/\varepsilon}\frac{1}{(1+k^2)^{1/2}}dk = \sinh^{-1}(k)\,|_{-l/\varepsilon}^{l/\varepsilon} = \log\left(l+\sqrt{l^2+\varepsilon^2}\right)-\log\left(-l+\sqrt{l^2+\varepsilon^2}\right),$$

$$\int_{-l/\varepsilon}^{l/\varepsilon}\frac{k}{(1+k^2)^{1/2}}dk = \left(1+k^2\right)^{1/2}|_{-l/\varepsilon}^{l/\varepsilon} = 0,$$

substituting the asymptotic series into the integral gives

$$
\begin{aligned}
&\mathbf{A}_{j,local}(p) \\
&= \frac{\Gamma_j}{4\pi}\Bigg[\mathbf{t}_j(s_\mathrm{f})\Big\{\log\left(l+\sqrt{l^2+\varepsilon^2}\right)-\log\left(-l+\sqrt{l^2+\varepsilon^2}\right) \\
&+\varepsilon\left(-\frac{2l}{\sqrt{\varepsilon^2+l^2}}+\log\left(l+\sqrt{l^2+\varepsilon^2}\right)-\log\left(-l+\sqrt{l^2+\varepsilon^2}\right)\right)\left(\frac{\cos\theta}{2}\frac{d^2x}{ds_j^2}_{|s_\mathrm{f}}+\frac{\sin\theta}{2}\frac{d^2y}{ds_j^2}_{|s_\mathrm{f}}\right)\Big\}+\mathrm{h.o.t}\Bigg].
\end{aligned}
$$

The first two terms are both in the direction of the unit tangent at the point p_f and when combined give rise to a $O(\log\varepsilon)$ term. The leading order behavior of $\mathbf{A}_j$ is therefore characterized by a vector field in the unit tangent direction with a $\log\varepsilon$ behavior, as $\varepsilon \to 0$. Note, however, that the next order field in $\mathbf{A}_j$, after the leading order, may be different from that predicted by $\mathbf{A}_{j,local}$ alone when one takes into account the contribution from $\mathbf{A}_{j,non-local}$.

Appendix C
Hamiltonian Function and Vector Field in the Half-space Model for $N_p = 2$ [151]

The expanded form of the Hamiltonian function H_{sym} of (3.54) for $N_p = 2$ is

$$
\begin{aligned}
H_{sym} \\
&= \frac{1}{2c}\left(L - \sum_{j=1}^{2} 2\Gamma_j y_j + 2R^2 \sum_{j=1}^{2} \Gamma_j \left(\frac{y_j}{x_j^2 + y_j^2}\right)\right)^2 \\
&+ \sum_{j=1}^{2} \Gamma_j^2 \left(\frac{1}{2\pi}\log(l_j^2 - R^2) - \frac{1}{4\pi}\log(l_j^4 + R^4 - 2R^2(x_j^2 - y_j^2)) + \frac{1}{2\pi}\log(4y_j)\right) \\
&+ 2\Gamma_1\Gamma_2 \left(\frac{1}{4\pi}\log(l_1^2 l_2^2 + R^4 - 2R^2(x_1x_2 + y_1y_2)) - \frac{1}{4\pi}\log(l_1^2 l_2^2 + R^4 - 2R^2(x_1x_2 - y_1y_2))\right. \\
&\qquad \left. + \frac{1}{4\pi}\log((x_1 - x_2)^2 + (y_1 + y_2)^2) - \frac{1}{4\pi}\log((x_1 - x_2)^2 + (y_1 - y_2)^2)\right)
\end{aligned}
$$

and the expanded form of the Hamiltonian vector field is

$$
\begin{aligned}
\frac{dx_1}{dt} \\
&= \left(-\frac{1}{c} + \frac{R^2}{c}\left(\frac{x_1^2 - y_1^2}{(x_1^2 + y_1^2)^2}\right)\right)\left(L - 2\Gamma_2 y_2 + 2R^2\Gamma_2\left(\frac{y_2}{x_2^2 + y_2^2}\right)\right) + \frac{2\Gamma_1}{c}y_1 - \frac{4\Gamma_1 R^2}{c}\frac{x_1^2 y_1}{(x_1^2 + y_1^2)^2} \\
&\quad + \frac{2R^4\Gamma_1}{c}\frac{y_1(x_1^2 - y_1^2)}{(x_1^2 + y_1^2)^3} + \Gamma_1\left(\frac{1}{2\pi}\frac{y_1}{l_1^2 - R^2} - \frac{1}{2\pi}\frac{(x_1^2 + y_1^2 + R^2)y_1}{l_1^4 + R^4 - 2R^2(x_1^2 - y_1^2)} + \frac{1}{4\pi y_1}\right) \\
&\quad + \frac{\Gamma_2}{4\pi}\left(\frac{2y_1 l_2^2 - 2R^2 y_2}{l_1^2 l_2^2 + R^4 - 2R^2(x_1x_2 + y_1y_2)} - \frac{2y_1 l_2^2 + 2R^2 y_2}{l_1^2 l_2^2 + R^4 - 2R^2(x_1x_2 - y_1y_2)}\right. \\
&\qquad \left. + \frac{2(y_1 + y_2)}{(x_1 - x_2)^2 + (y_1 + y_2)^2} - \frac{2(y_1 - y_2)}{(x_1 - x_2)^2 + (y_1 - y_2)^2}\right),
\end{aligned}
$$

B. N. Shashikanth, *Dynamically Coupled Rigid Body-Fluid Flow Systems*,
https://doi.org/10.1007/978-3-030-82646-8

$$
\begin{aligned}
\frac{dy_1}{dt}
&= \frac{2R^2}{c}\frac{x_1y_1}{(x_1^2+y_1^2)^2}\left(L-2\Gamma_2y_2+2R^2\Gamma_2\left(\frac{y_2}{x_2^2+y_2^2}\right)\right)-\frac{4R^2\Gamma_1}{c}\frac{x_1y_1^2}{(x_1^2+y_1^2)^2}+\frac{4R^4\Gamma_1}{c}\frac{x_1y_1^2}{(x_1^2+y_1^2)^3}\\
&\quad +\Gamma_1\left(-\frac{1}{2\pi}\frac{x_1}{l_1^2-R^2}+\frac{1}{2\pi}\frac{(x_1^2+y_1^2-R^2)x_1}{l_1^4+R^4-2R^2(x_1^2-y_1^2)}\right)\\
&\quad -\frac{\Gamma_2}{4\pi}\left(\frac{2x_1l_2^2-2R^2x_2}{l_1^2l_2^2+R^4-2R^2(x_1x_2+y_1y_2)}-\frac{2x_1l_2^2-2R^2x_2}{l_1^2l_2^2+R^4-2R^2(x_1x_2-y_1y_2)}\right.\\
&\quad \left.+\frac{2(x_1-x_2)}{(x_1-x_2)^2+(y_1+y_2)^2}-\frac{2(x_1-x_2)}{(x_1-x_2)^2+(y_1-y_2)^2}\right)
\end{aligned}
$$

$$
\begin{aligned}
\frac{dx_2}{dt}
&= \left(-\frac{1}{c}+\frac{R^2}{c}\left(\frac{x_2^2-y_2^2}{(x_2^2+y_2^2)^2}\right)\right)\left(L-2\Gamma_1y_1+2R^2\Gamma_1\left(\frac{y_1}{x_1^2+y_1^2}\right)\right)+\frac{2\Gamma_2}{c}y_2-\frac{4\Gamma_2R^2}{c}\frac{x_2^2y_2}{(x_2^2+y_2^2)^2}\\
&\quad +\frac{2R^4\Gamma_2}{c}\frac{y_2(x_2^2-y_2^2)}{(x_2^2+y_2^2)^3}+\Gamma_2\left(\frac{1}{2\pi}\frac{y_2}{l_2^2-R^2}-\frac{1}{2\pi}\frac{(x_2^2+y_2^2+R^2)y_2}{l_2^4+R^4-2R^2(x_2^2-y_2^2)}+\frac{1}{4\pi y_2}\right)\\
&\quad +\frac{\Gamma_1}{4\pi}\left(\frac{2y_2l_1^2-2R^2y_1}{l_1^2l_2^2+R^4-2R^2(x_1x_2+y_1y_2)}-\frac{2y_2l_1^2+2R^2y_1}{l_1^2l_2^2+R^4-2R^2(x_1x_2-y_1y_2)}\right.\\
&\quad \left.+\frac{2(y_1+y_2)}{(x_1-x_2)^2+(y_1+y_2)^2}-\frac{2(y_2-y_1)}{(x_1-x_2)^2+(y_1-y_2)^2}\right),
\end{aligned}
$$

$$
\begin{aligned}
\frac{dy_2}{dt}
&= \frac{2R^2}{c}\frac{x_2y_2}{(x_2^2+y_2^2)^2}\left(L-2\Gamma_1y_1+2R^2\Gamma_1\left(\frac{y_1}{x_1^2+y_1^2}\right)\right)-\frac{4R^2\Gamma_2}{c}\frac{x_2y_2^2}{(x_2^2+y_2^2)^2}+\frac{4R^4\Gamma_2}{c}\frac{x_2y_2^2}{(x_2^2+y_2^2)^3}\\
&\quad +\Gamma_2\left(-\frac{1}{2\pi}\frac{x_2}{l_2^2-R^2}+\frac{1}{2\pi}\frac{(x_2^2+y_2^2-R^2)x_2}{l_2^4+R^4-2R^2(x_2^2-y_2^2)}\right)\\
&\quad -\frac{\Gamma_1}{4\pi}\left(\frac{2x_2l_1^2-2R^2x_1}{l_1^2l_2^2+R^4-2R^2(x_1x_2+y_1y_2)}-\frac{2x_2l_1^2-2R^2x_1}{l_1^2l_2^2+R^4-2R^2(x_1x_2-y_1y_2)}\right.\\
&\quad \left.+\frac{2(x_2-x_1)}{(x_1-x_2)^2+(y_1+y_2)^2}-\frac{2(x_2-x_1)}{(x_1-x_2)^2+(y_1-y_2)^2}\right)
\end{aligned}
$$

References

1. Abdel Moneim, A. F.: Motion of a system of rigid bodies through an ideal incompressible fluid. J. Ship Research, **20**, 152–159 (1976).
2. Abraham, R., Marsden, J. E.: Foundations of Mechanics, second edition. AMS Chelsea Publishing, Providence, Rhode Island (1978).
3. Abraham, R., Marsden, J. E., Ratiu, T. S.: Manifolds, Tensor Analysis and Applications, Second Edition, Applied Mathematical Sciences **75**, Springer-Verlag, New York (1988).
4. Acheson, D.J.: Instability of vortex leapfrogging. Eur. J. Phys., **21**, 269–273 (2000).
5. Alemi Ardakani, H., Bridges, T. J., Gay-Balmaz, F., Huang, Y. H., Tronci, C.: A variational principle for fluid sloshing with vorticity, dynamically coupled to vessel motion. Proc. Roy. Soc. A, Mathematical, Physical and Engineering Sciences (2019). https://doi.org/10.1098/rspa.2018.0642
6. Allen, J. J., Jouanne, Y., Shashikanth, B. N.: Vortex interaction with a moving sphere. J. Fluid Mech., **587**, 337–346 (2007).
7. Aref, H., Jones, S. W.: Chaotic motion of a solid through ideal fluid. Phys. Fluids A, **5**, 3026–3028 (1993).
8. Arnold, V. I.: Sur la géométrie différentielle des groupes de Lie de dimension infinie et ses applications à l'hydrodynamique des fluides parfaits. Annales de l'institut Fourier, **16**, 319–361 (1966).
9. Arnold, V. I.: On an a priori estimate in the theory of hydrodynamic stability, Izv. Vyssh. Uchebn. Zaved. Mat., **5**, 3–5 (1966).
10. Arnold, V. I.: The Hamiltonian nature of the Euler equation in the dynamics of a rigid body and of an ideal fluid. Uspekhi Mat. Nauk. **24**, 225–226, (1969).
11. Arnold, V.I: Mathematical methods of classical mechanics, second edition. Springer-Verlag, New York (1989).
12. Arnold, V. I., Khesin, B.: Topological Methods in Hydrodynamics, Appl. Math. Sciences, **125**. Springer-Verlag (1998).
13. Athanassoulis, G. A.: Analytical dynamics of wave body interaction. Mathematical techniques for water waves (ed. B. N. Mandal), Advances in Fluid Mechanics, **8**, Computational Mechanics Publications, Southampton, 79–154,

B. N. Shashikanth, *Dynamically Coupled Rigid Body-Fluid Flow Systems*,
https://doi.org/10.1007/978-3-030-82646-8

(1997) Available from: https://www.researchgate.net/publication/266301554_Analytical_dynamics_of_wave-body_interaction.
14. Athanassoulis, G. A., Loukakis, T. A.: Lagrangian expressions of the hydrodynamic forces acting on a rigid body in the presence of a free surface. J. Ship Research, **29**, 12–22 (1985).
15. Batchelor, G. K.: An Introduction to Fluid Dynamics. Cambridge University Press (1967).
16. Birkhoff, G.: Hydrodynamics: A study in logic, fact and similitude, revised edition. Princeton University Press, Princeton (1960).
17. Blackmore, D., Brøns, M., Goullet, A.: A coaxial vortex ring model for vortex breakdown. Physica D: Nonlinear Phenomena, **237**, 2817–2844 (2008).
18. Blackmore, D., Champanerkar, J., Wang, C.:. A generalized Poincaré-Birkhoff theorem with applications to coaxial vortex ring motion. Discrete and Continuous Dynamical Systems-B, **5**, 15–33 (2005).
19. Blackmore, D., Knio, O.: KAM theory analysis of the dynamics of three coaxial vortex rings. Physica D: Nonlinear Phenomena, **140**, 321–348 (2000).
20. Blackmore, D., Knio, O.: Transition from quasiperiodicity to chaos for three coaxial vortex rings. Z. angew. Math. Mech., **80**, 173–176 (2000).
21. Blackmore, D., Knio, O.: Hamiltonian formulation of the dynamics of interaction between a vortex ring and a rigid sphere. European Congress on Computational Methods in Applied Sciences and Engineering ECCOMAS 2000, Barcelona, 11–14 September (2000).
22. Borisov, A.V., Kilin, A. A., Mamaev, I. S., Tenenev, V. A.: The dynamics of vortex rings: leapfrogging in an ideal and viscous fluid. Fluid Dyn. Res., **46**, 031415 (2014).
23. Borisov, A. V., Mamaev, I. S.:, An integrability of the problem on motion of cylinder and vortex in the ideal fluid. Reg. Chaotic Dyn., **8**, 163–166 (2003).
24. Borisov, A. V., Mamaev, I. S., Ramodanov, S. M.: Motion of a circular cylinder and n point vortices in a perfect fluid. Reg. Chaotic Dyn., **8**, 449–462 (2003).
25. Borisov, A. V., Mamaev, I. S., Ramodanov, S. M.: Dynamic interaction of point vortices and a two-dimensional cylinder. J. Math. Phys, **48**, 1-1 to 1–9 (2007).
26. Brooke Benjamin, T., Bridges, T. J.: Reappraisal of the Kelvin-Helmholtz problem. Part 1. Hamiltonian structure. J. Fluid Mech., **333**, 301–325 (1997).
27. Brooke Benjamin, T., Olver, P. J.: Hamiltonian structure, symmetries and conservation laws for water waves. J. Fluid Mech., **125**, 137–185 (1982).
28. Cai, J., Liu, F., Luo, S.: Stability of symmetric vortices in two dimensions and over three-dimensional slender conical bodies. J. Fluid Mech., **480**, 65–94 (2003).
29. Callegari, A. J., Ting. L.: Motion of a curved vortex filament with decaying vortical core and axial velocity, *SIAM Journal on Applied Mathematics*, **35**, 148–175 (1978).
30. Cantor, M.: Groups of diffeomorphisms of $\mathbb{R}^n$ and the flow of a perfect fluid. Bull. Amer. Math. Soc., **81**, 205–208 (1975).

31. Cendra, H., Marsden, J. E.: Geometric mechanics and the dynamics of asteroid pairs. Dynamical Systems. An International Journal, **20**, 3–21 (2005).
32. Cendra, H., Marsden, J. E. and Ratiu, T. S.: Lagrangian reduction by stages. Memoirs, American Mathematical Society, **152**. Providence, R.I. (2001).
33. Chang, C. C., Chu, C. C., Liu, C. C., Chang, C. C., Lee, S. T.: Flow induced by a pair of line vortices moving against a circular-cylinder. J. Phys. Soc. Japan, **64**, 1557–1578 (1995).
34. Chapman, D. M. F.: Ideal vortex motion in two dimensions: Symmetries and conservation laws. J. Math. Phys., **19**, 1988–1992 (1978).
35. Clebsch, A.: Ueber die Bewegung eines Körpers in einer Flüssigkeit. Mathematischen Annalen, **3**, 238–262 (1870).
36. Crowdy, D., Marshall, J.: Analytical formulae for the Kirchhoff–Routh path function in multiply connected domains. Proc. R. Soc. A, **461**, 2477–2501 (2005).
37. Crowdy, D. G., Marshall, J. S.: The motion of a point vortex around multiple circular islands. Phys. Fluids, **17**, 056602, (2005).
38. do Carmo, M. P.: Differential Forms and Applications, Universitext, Springer-Verlag, Berlin (1994).
39. Dhanak, M. R.: Interaction between a vortex filament and an approaching rigid sphere. J. Fluid Mech., **110**, 129–147 (1981).
40. Dyson, F.: The potential of an anchor ring. Phil. Trans. Roy. Soc. London Ser. A, **184**, 43–95 (1893).
41. Dyson, F.: The potential of an anchor ring-Part II. Phil. Trans. Roy. Soc. London Ser. A, **184**, 1041–1106 (1893).
42. Ebin, D. G., Marsden, J. E.: Groups of diffeomorphisms and the motion of an incompressible fluid. Ann. Math., **92**, 102–163 (1970).
43. Elcrat, A., Fornberg, B., Horn, M., Miller, K: Some steady vortex flows past a circular cylinder. J. Fluid Mech., **409**, 13–27 (2000).
44. Faddeev, L. D., Takhtajan, L. A.: Hamiltonian methods in the theory of solitons. Classics in Mathematics, Springer (2007).
45. Flucher, M., Gustafsson, B.: Vortex motion in two-dimensional hydrodynamics. https://people.kth.se/gbjorn/vortex.pdf.
46. Föppl, L.: Wirbelbewegung hinter einem Kreiszylinder, Sitz. K. Bäyr Akad. Wiss, **1**, 7–18 (1913). (Also: "Vortex motion behind a circular cylinder," NASA TM, **77015**).
47. Galper, A., Miloh, T.: Generalized Kirchhoff equations for a deformable body moving in a weakly non-uniform flow field. Proc. R. Soc. London A, **446**, 169–193 (1994).
48. Galper, A., Miloh, T.: Dynamic equations of motion for a rigid or deformable body in an arbitrary nonuniform potential flow-field. J. Fluid Mech., **295**, 91–120 (1995).
49. Galper, A., Miloh, T.: Hydrodynamics and stability of a deformable body moving in the proximity of interfaces. Phys. Fluids, **11**, 795–806 (1999).
50. Gardner, C. S., Greene, J. M., Kruskal, M. D., Miura, R. M.: Method for solving the Korteweg-de Vries equation. Phys, Rev. Lett., **19**, 1095–1097 (1967).

51. Goldstein, H.: Classical Mechanics. Addison-Wesley, Massachusetts (1950).
52. Gröbli,W.: Spezielle Probleme über die Bewegung geradliniger paralleler Wirbelfäden. PhD Thesis, University of Göttingen. Zürcher und Furrer, Zürich (1877). Also in Vierteljschr. Nat. Ges. Zürich **22**, 37–82, 129–168 (1877).
53. Gurzhii, A. A., Konstantinov, M. Yu., V. V. Meleshko, V. V.: Interaction of coaxial vortex rings in an ideal fluid, *Fluid Dynam.*, **23**, 224–229, (1988) (translated from Izv. Akad. Nauk SSSR Mekh. Zhidk. Gaza, OVYR, **2**,78–84 (Russian) (1988)).
54. Guzman, J. E. V., Kamp, L. P. J., van Heijst, G. J. F.: Vortex dipole collision with a sliding wall. Fluid Dyn. Research, **45**, Article Number: 045501 (2013).
55. Hartmann, D., Schneiders, L., Schröder, W., Shashikanth, B. N.: On the interaction of a vortex pair with a freely moving cylinder. AIAA 2010-4749, 40th Fluid Dynamics Conference and Exhibit, Chicago, Illinois, 28 June–1 July (2010).
56. Hartmann, D., Schröder, Shashikanth, B. N.: Non-invasive determination of external forces in vortex-pair-cylinder interactions. Phys. Fluids, **24**, 061903 (2012).
57. Harvey, J. K., Perry, F. J.: Flowfield produced by trailing vortices in the vicinity of the ground. AIAA J., **9**, 1659–1660 (1971).
58. Hasimoto, H.: A soliton on a vortex filament. J. Fluid Mech., **51**, 477–485 (1972).
59. Helmholtz, H.: Über Integrale der hydrodynamischen Gleichungen, welche den Wirbelbewegungen entsprechen. J. Reine Angew. Math., **55**, 25–55 (1858).
60. Helmholtz, H.: On integrals of the hydrodynamical equations, which express vortex-motion. Phil. Mag. (Ser. 4),**33**, 485–510 (1867).
61. Hernández-Garduño, A., Lacomba, E. A.: Collisions of four point vortices in the plane. arXiv preprint math-ph/0609016 (2006).
62. Hernández-Garduño, A., Shashikanth, B. N.: Reconstruction phases in the planar three- and four-vortex problems. Nonlinearity, **31**, 783–814 (2018).
63. Hicks, W.M.: On the mutual threading of vortex rings. Proc. Roy. Soc. A, **102**, 111–131 (1922).
64. Hill, D.: Vortex dynamics in wake models. Ph.D thesis, Caltech, USA (1998).
65. Hirasaki, G. J., Hellums, J. D.: A general formulation of the boundary conditions on the vector potential in three-dimensional hydrodynamics. Quart. Appl. Math., **26**, 331–342 (1968).
66. Holm, D. D.: Geometric Mechanics, Part I: Dynamics and Symmetry. Geometric mechanics, Part II: Rotating, Translating and Rolling. Imperial Collge Press (2008).
67. Holmes, P., Jenkins, J., Leonard, N. E.: Dynamics of the Kirchhoff equations I: Coincident centers of gravity and buoyancy. Physica D, **118**, 311–342 (1998).
68. Homa, J., Lucas, M., Rockwell, D.: Interaction of imulsively generated vortex pairs with bodies. J. Fluid Mech., **197**, 571–594 (1988).

69. Horowitz, M., Williamson, C. H. K.: Critical mass and a new periodic four-ring vortex wake mode for freely rising and falling spheres. Phys. Fluids, **20**, 101701 (2008).
70. Horowitz, M., Williamson, C. H. K.: The effect of Reynolds number on the dynamics and wakes of freely rising and falling spheres. J. Fluid Mech., **651**, 251–294 (2010).
71. Horowitz, M., Williamson, C. H. K.: Vortex-induced vibration of a rising and falling cylinder. J. Fluid Mech., **662**, 352–383 (2010).
72. Howe, M. S.: On the force and moment on a body in an incompressible fluid, with application to rigid bodies and bubbles at high and low Reynolds numbers. Quart. J. Mech. appl. Math., **48**, 401–426 (1995).
73. Hussein, A. A., Taha, H. E., Ragab, S., Hajj, M. R.: A variational approach for the dynamics of unsteady point vortices. Aerospace Science and Technology, **78**, 559–568 (2018).
74. Ilinskii, Y. A., Hamilton, M. F., Zabolotskaya, E, A.: Bubble interaction dynamics in Lagrangian and Hamiltonian mechanics. The Journal of the Acoustical Society of America, **121**, 786–795 (2007).
75. John, F.: On the motion of floating bodies. I. Comm. Pure App. Math., **2**, 13–57 (1949).
76. John, F.: On the motion of floating bodies. II. Simple harmonic motions. Comm. Pure App. Math., **3**, 45–101 (1950).
77. Johnson, R. S.: A Modern Introduction to the Mathematical Theory of Water Waves. Cambridge Texts in Applied Mathematics, Cambridge University Press (1997).
78. Jonkman, J. M.: Dynamics modeling and loads analysis of an offshore floating wind turbine, Technical Report, NREL/TP-500-41958 (2007).
79. Kanso, E., Marsden, J. E., Rowley, C., Melli-Huber, J. B.: Locomotion of articulated bodies in a perfect fluid. J. Nonlinear Sci., **15**, 255–289, (2005).
80. Kanso, E., Oskouei, B. G.: Stability of a coupled body-vortex system. J. Fluid Mech., **600**, 77–94 (2008).
81. Kirchhoff, G.: Ueber die Bewegung eines Rotationskörpers in einer Flüssigkeit. Journal für die reine und angewandte Mathematik (Crelle's Journal), **71**, 237–262 (1869).
82. Kirchhoff, G.: Vorlesungen über Mathematische Physik von Dr. Gustav Kirchhoff, Mechanik, Zwanzigste Vorlesung. B. G. Teubner, Leipzig (1876).
83. Klein R., Majda, A. J., Damodaran, K.: Simplified equations for the interaction of nearly parallel vortex filaments, *J. Fluid Mech.*, **288**, 201–248 (1995).
84. Knio, O. M., Ting, L.: Vortical flow outside a body and sound generation. SIAM J. Appl. Math., **57**(4), 972–981 (1997).
85. Koiller, J.: Note on coupled motions of vortices and rigid bodies. Phys. Lett. A, **120**, 391–395 (1987).
86. Koiller, J., Carvalho, S. P.: Non-integrability of the 4-vortex system: analytical proof. Comm. Math. Phys., **120**, 643–652 (1989).
87. Konstantinov, M.: Chaotic phenomena in the interaction of vortex rings. Phys. Fluids **6**, 1752–1767 (1994).

88. Korotkin, A. I.: Added Masses of Ship Structures. Springer, Netherlands (2009).
89. Kozlov, V. V., Oniščenko, D. A.: Nonintegrability of Kirchhoff's equations. Soviet Math. Dokl., **26**, 495–498 (1982).
90. Kudela, H.: Self-similar collapse of n point vortices. J. Nonlinear Sci. https://doi.org/10.1007/s00332-014-9207-8.
91. Kuznetsov, E. A., Ruban, V. P.: Cherenkov interaction of vortices with a free surface. J. Exp. Theor. Phys., **88**, 492–505 (1999).
92. Lamb, H.: Hydrodynamics, 6th edition. Dover, New York (1932).
93. Lannes, D.: On the dynamics of floating structures, *Ann. PDE*, **3**(11), https://doi.org/10.1007/s40818-017-0029-5, (2017).
94. Larmor, J.: Electro-magnetic and other images in spheres and planes. Quart. J. Pure and Appl. Math, **23**, 94–101 (1889).
95. Leonard, N. E.: Stability of a bottom-heavy underwater vehicle. Automatica, **33**, 331–346 (1997).
96. Lewis, D., Marsden, J., Montgomery, R, Ratiu, T.: The Hamiltonian structure for dynamic free boundary problems. Physica D, **18**, 391–404 (1986).
97. Lewis, T.C.: On the images of vortices in a spherical vessel. Quart. J. Pure and Appl. Math, **16**, 338–347 (1879).
98. Lighthill, M. J.: The image system of a vortex element in a rigid body. Proc. Cambr. Phil. Soc., **52**, 317–321 (1956).
99. Lighthill, M. J.: Hydromechanics of aquatic animal propulsion. Annu. Rev. Fluid Mech., **1**, 413–45 (1969).
100. Lim, T.T., Nickels, T. B.: Instability and reconnection in the head-on collision of two vortex rings. Nature, **357**, 225–227 (1992).
101. Lin, C. C.: On the motion of vortices in two dimensions-I and II. Proc. Nat. Acad. Sci., **27**, 570–577 (1941).
102. Lorenz, E. N.: Deterministic nonperiodic flow. J. Atmos. Sci., **20**, 130–141 (1963).
103. Love, A. E. H.: On the motion of paired vortices with a common axis. Proc. Lond. Math. Soc., **s1-25**, 185–194 (1893).
104. Lundgren, T., Koumoutsakos, P.: On the generation of vorticity at a free surface. J. Fluid Mech., **382**, 351–366 (1999).
105. Marsden, J. E.: Lectures on Mechanics, London Mathematical Society Lecture Note Series, Cambridge University Press (1992).
106. Marsden, J.E., Misiolek, G., Perlmutter, M. and Ratius, T. S.: Hamiltonian reduction by stages. Lecture Notes in Mathematics, **1913**. Springer-Verlag Berlin Heidelberg (2007).
107. Marsden, J. E., Montgomery, R., Ratiu, T. S.: Reduction, symmetry and phases in mechanics Memoirs, Amer. Math. Soc., **436**, Providence, RI (1990).
108. Marsden, J. E., Ratiu, T. S.: Reduction of Poisson manifolds. Lett. Math. Phys., **11**, 161–170 (1986).
109. Marsden, J. E., Ratiu, T.S.: Introduction to Mechanics and Symmetry, Second Edition. Springer-Verlag, New York (1999).

110. Marsden, J. E., Weinstein, A.: Reduction of symplectic manifolds with symmetry. Rep. Math. Phys. **5**, 121–130 (1974).
111. Marsden, J. E., Weinstein, A.: Coadjoint orbits, vortices and Clebsch variables for incompressible fluids. Physica D, **7**, 305–323 (1983).
112. Meleshko, V.V., Aref, H.: A bibliography of vortex dynamics 1858–1956. Advances in Applied Mechanics, **41**, 197–292 (2007).
113. Michelin, S. H. R.: Falling, flapping, flying, swimming,... : high-Re fluid-solid interactions with vortex shedding. PhD. Thesis, University of California, San Diego (2009).
114. Miloh, T.: Hamilton's principle, Lagrange's method, and ship motion theory. J. Ship Research, **28**, 229–237 (1984).
115. Miloh, T.: A note on impulsive sphere motion beneath a free-surface. J. Engg. Math., **41**, 1–11 (2001).
116. Milne-Thomson, L. M.: Hydrodynamical images. Mathematical Proceedings of the Cambridge Philosophical Society, **36**, 246–247 (1940).
117. Milne-Thomson, L. M.: Theoretical Hydrodynamics, 5th edition. Dover, New York (1996).
118. Möhring, W.: Sound radiation by two elliptic vortex rings. J. Sound Vib., **140**, 155–162 (1990).
119. Moura, M. N., Vasconcelos, G. L.: Four-vortex motion around a circular cylinder. Phys. Fluids, **25**, 073601 (2013).
120. Newton, P. K.: The N-Vortex Problem: Analytical Techniques (Applied Mathematical Sciences), **145**, Springer-Berlin.
121. Novikov, E. A., Sedov, Yu. B.: Vortex collapse. Sov. Phys. JETP, **50**, 297–301 (1979).
122. Novikov, S. P., Shmel'tser, I.: Periodic solutions of the Kirchhoff equations for the free motion of a rigid body in a fluid and the extended Lyusternik-Shnirel'man-Morse theory. I. Funktsional Anal. i Prilozhen., **15**, 54–66 (1981). Available online from http://www.mi.ras.ru/~snovikov/69.pdf
123. Orlandi, P.: Vortex dipoles impinging on circular cylinders. Phys. Fluids A–Fluid Dynamics, **5**, 2196–2206 (1993).
124. Orlandi, P., Verzicco R.: Vortex rings impinging on walls–axisymmetric and 3-dimensional simulations. J. Fluid Mechanics, **256**, 615–646 (1993).
125. Oshima, Y: Head-on collision of two vortex rings. J. Phys. Soc. Japan, **44**, 328–331 (1978).
126. Pedlosky, J.: Geophysical fluid dynamics. Second edition, Springer, New York (1987).
127. Pedrizzetti, G: Close interaction between a vortex filament and a rigid sphere. J. Fluid Mech., **245**, 701–722, (1992).
128. Perelomov, A. M.: A few remarks about integrability of the equations of a rigid body in an ideal fluid. Phys. Lett., **80A**, 156–158 (1980).
129. Peterson. S. D., Porfiri, M.: Impact of a vortex dipole with a semi-infinite rigid plate. Phys. Fluids, **25**(9), Article Number: 093103 (2013).

130. Poincaré, H.: Théorie des tourbillons. Leçons profesées pendant le deuxième semestre 1891-92, G. Carre, Paris (1893) (available for online browsing at http://digital.library.cornell.edu/cgi/t/text/text-idx?c=math;idno=01800002).
131. Protas, B.: Linear feedback stabilization of laminar vortex shedding based on a point vortex model. Phys. Fluids, **16**, 4473–4488, (2004).
132. Quartapelle, L., Napolitano, M.: Force and moment in incompressible flows. AIAA Journal, **21**, 911–913 (1983).
133. Ramodanov, S. M: Motion of a circular cylinder and *N* point vortices in a perfect fluid. Reg. Chaotic Dyn., **7**, 291–298 (2002).
134. Ricca, R. L.: Rediscovery of DaRios equations. Nature, **352**, 561–562 (1991).
135. Ricca, R. L.: The contributions of Da Rios and Levi-Civita to asymptotic potential theory and vortex filament dynamics. Fluid Dyn. Res., **18**, 245–268 (1996).
136. Rivoalen, E., Huberson, S., Knio, O. M.: Sound generation by the interaction of a vortex ring with a rigid sphere. ESAIM Proceedings, Vol. 7, 369–377, Third International Workshop on Vortex Flows and Related Numerical Methods (1999). http://www.emath.fr/proc/Vol.7/
137. Roberts, P. H., Donnelly, R. J.: Dynamics of vortex rings. Phys. Lett., **31A**, 137–138 (1970).
138. Rockwell, D.: Vortex-body interactions. Ann. Rev. Fluid Mechanics, **30**, 199–229 (1998).
139. Roenby, J.: Chaos and integrability in ideal body-fluid interactions. Ph.D thesis, Technical University of Denmark, Lyngby, Denmark (2011).
140. Roenby, J., Aref, H.: Chaos in body-vortex interactions. Proc. R. Soc. A https://doi.org/10.1098/rspa.2009.0619 (2010).
141. Rood, E. P.: Vorticity interactions with a free surface. Fluid Vortices (ed. S. I. Green), Fluid Mechanics and Its Applications, **30**, Springer, Dordrecht, 687–730 (1995).
142. Rouhi, A., Wright, J.: Hamiltonian formulation for the motion of vortices in the presence of a free surface for ideal flow. Phys. Rev E, **48**, 1850–1865 (1993).
143. Rowley, C., Marsden, J. E.: Variational integrators for degenerate Lagrangians, with applications to point vortices. Proc. 41st IEEE Conf. Decision and Control, Las Vegas, Nevada, 1521–1527 (2002).
144. Rubanovskii, V. N.: Quadratic integrals of the equations of motion of a rigid body in a liquid. J. Appl. Math. Mech., **52**, 313–322 (1988).
145. Ryu, K.-W, Lee, D.-J.: Sound radiation from elliptic vortex rings: evolution and interaction. J. Sound Vib., **200**, 281–301 (1997).
146. Ryu, K. W, Lee, D. J.: Interaction between a vortex ring and a rigid sphere. Eur. J. Mech. B/Fluids, **16**, 645–664 (1997).
147. Saffman, P. G.: Vortex Dynamics, Cambridge Monographs on Mechanics and Applied Mathematics. Cambridge University Press (1992).
148. Salvesen, N., Tuck, E. O., Faltinsen, O.: Ship motions and sea loads. Transactions of the Society of Naval Architects and Marine Engineers,

6, 1–30 (1970). Available from: www.marinecontrol.org/References/papers/SalvesenTuckFaltinsen1970.PDF

149. Seath, D. D.: Equilibrium vortex positions. J. Spacecraft and Rockets, **8**, 72–75 (1971).
150. Shashikanth, B. N.: Poisson brackets for the dynamically interacting system of a 2D rigid boundary and N point vortices: The case of arbitrary smooth cylinder shapes. Reg. Chaotic Dyn., **10**, 1–14 (2005).
151. Shashikanth, B. N.: Symmetric pairs of point vortices interacting with a neutrally buoyant 2D circular cylinder. Phys. Fluids, **18**, 127103 (2006).
152. Shashikanth, B. N.: Symmetry reduction and control of the dynamics of a 2D rigid circular cylinder and a point vortex: vortex capture and scattering. Euro. J. Control, **6**, 1–15, (2007).
153. Shashikanth, B. N.: Kirchhoff's equations of motion via a constrained Zakharov system. J. Geom. Mech., **8**, 461–485 (2016).
154. Shashikanth, B. N.: Poisson brackets for the dynamically coupled system of a free boundary and a neutrally buoyant rigid body in a body-fixed frame. J. Geom. Mech., **12**, 25–52 (2020).
155. Shashikanth, B. N., Marsden, J. E.: Leapfrogging vortex rings: Hamiltonian structure, geometric phases and discrete reduction. Fluid Dyn. Res., **33**, 333–356 (2003).
156. Shashikanth, B. N., Marsden, J. E., Burdick, J. W., Kelly, S. D.: The Hamiltonian structure of a 2-D rigid cylinder interacting dynamically with N point vortices. Phys. Fluids, **14**, 1214–1227 (2002).
157. Shashikanth, B. N., Sheshmani, A., Kelly, S. D., Marsden, J. E.: Hamiltonian Structure for a Neutrally Buoyant Rigid Body Interacting with N Vortex Rings of Arbitrary Shape: The Case of Arbitrary Smooth Body Shape. Theoretical and Computational Fluid Dynamics, **22**, 37–64 (2008).
158. Shashikanth, B. N., Shashmani, A., Kelly, S. D., Wei, M. J.: Hamiltonian structure and dynamics of a neutrally buoyant rigid sphere interacting with thin vortex rings. J. Mathematical Fluid Mechanics, **12**, 335–353 (2010).
159. Souriau, J. M.: Structure of Dynamical Systems: A Symplectic View of Physics. Progress in Mathematics, **149**. Birkhäuser (1997).
160. Synge, J. L.: On the motion of three vortices. Can. J. Math., **1**, 257–270 (1949).
161. Tait, P. G.: Prof.Kirchhoff. Nature, **36**, 321–354 (1887).
162. Tang, S., Aubry, N.: On the symmetry breaking instability leading to vortex shedding. Phys. Fluids, **9**, 2550–2561 (1997).
163. Thomson, W., Tait, P.G.: Treatise on Natural Philosophy. Volume 1, Part 1, New Edition. Cambridge: At the University Press (1879).
164. Thorpe, J. A.: Elementary Topics in Differential Geometry. Undergraduate Texts in Mathematics. Springer-Verlag (1979).
165. Ting, L.: On the application of integral invariants and decay laws of vorticity distributions. J. Fluid Mech., **127**, 497–506 (1983).
166. Ting, L., Klein, R.: Viscous Vortical Flows. Lecture Notes in Physics, **374**. Springer, Berlin (1991).

167. Topøh, L., Aref, H.: Instability of vortex pair leapfrogging. Phys. Fluids., **25**, 014107 (2013).
168. Triantafyllou, M. S., Triantafyllou, G. S., Yue, D. K. P.: Hydrodynamics of fishlike swimming. Annu. Rev. Fluid Mech., **32**, 33–53 (2000).
169. Ursell, F.: On the heaving motion of a circular cylinder on the surface of a fluid. Quart. J. Mech. App. Math., **II**, 218–231 (1949).
170. Ursell, F.: Short surface waves due to an oscillating immersed body. Proc. R. Soc. Lond. A, **220** (1953), 90–103 (1953).
171. van Daalen, E.: Numerical and Theoretical Studies of Water Waves and Floating Bodies, PhD Thesis, University of Twente (1993).
172. van Daalen, E. F. G., van Groesen, E., Zandbergen, P. J.: A Hamiltonian formulation for nonlinear wave-body interactions. Eighth International Workshop on Water Waves and Floating Bodies 23–26 May 1993, St John's, Newfoundland, Canada, 159–163 (1993).
173. Vankerschaver, J., Kanso, E., Marsden, J.: The geometry and dynamics of interacting rigid bodies and point vortices, J. Geom. Mech., **1**, 223–266 (2009).
174. Vankerschaver, J., Kanso, E., Marsden, J. E.: The dynamics of a rigid body in potential ow with circulation. Regular and Chaotic Dynamics, **15**, 606–629 (2010).
175. Velasco-Fuentes, O. U.: On the topology of vortex lines and tubes. J. Fluid Mech., **584**, 147–156 (2007).
176. Verzicco, R., Flor, J. B., van Heijst, G. J. F., Orlandi, P.: Numerical and experimental-study of the interaction between a vortex dipole and a circular cylinder. Experiments in Fluids, **18**, 153–163 (1995).
177. Voinov, O. V., A. M. Golovin, A. M.: Lagrange equations for a system of bubbles of varying radii in a liquid of small viscosity. Fluid Dynamics, **5**, 458–464 (1970). (Originally: Izv. AN SSSR, Mekhanika Zhidkosti i Gaza, Vol. 5, No. 3, pp. 117–123, 1970.)
178. Voinov, V. V., Voinov, O. V., Petrov, A. G.: Hydrodynamic interaction between bodies in a perfect incompressible fluid and their motion in non-uniform streams. J. App. Math. Mech (PMM), **37**, 680–689 (1973).
179. Wakelin, S.L., Riley, N.: Vortex ring interactions II. Inviscid models. Quart. J. Mech. App. Math., **49**, 287–309 (1996).
180. Warner, F. W.: Foundations of Differentiable Manifolds and Lie Groups. Graduate Texts in Mathematics, **94**. Springer-Verlag (1994).
181. Wehausen, J. V.: The motion of floating bodies. Ann. Rev. Fluid Mech., **3**, 237–268 (1971).
182. Weihs, D., Boasson, M.: Multiple equilibrium vortex positions in symmetric shedding from slender bodies. AIAA Journal, **17**, 213–214 (1979).
183. Weinstein, A.: The local structure of Poisson manifolds. J. Diff. Geometry, **18**, 523–557.
184. Weiss, P.: On hydrodynamical images: Arbitrary irrotational flow disturbed by a body. Proc. Cambridge Phil. Soc., **40**, 259–261 (1944).
185. Weissmann, S.: Hamiltonian dynamics of several rigid bodies interacting with point vortices. J. Nonlinear Sci., **24**, 359–382 (2014).

186. Wells, J. C.: A geometrical interpretation of force on a translating body in rotational flow. Phys. Fluids, **8**, 442–450 (1996).
187. Wells, J. C.: Calculating the impulse and angular impulse of image vorticity by integrals over free vorticity. Proc. R. Soc. Lond. A, **454**, 2791–2809 (1998).
188. Wu, G. X., Taylor, R. E.: The coupled finite element and boundary element analysis of nonlinear interactions between waves and bodies. Ocean Engg., **30**, 387–400 (2003).
189. Wu, J. C.: Theory for aerodynamic force and moment in viscous flows. AIAA Journal, **19**, 432–441 (1981).
190. Wu, J-Z.: A theory of three-dimensional interfacial vorticity dynamics. Phys. Fluids, **7**, 2375–2395 (1995).
191. Wu, J.-Z., Wu, J.-C.: Interactions between a solid surface and a viscous compressible flow field. J. Fluid Mech., **254**, 183–211 (1993).
192. Zakharov, V. E.: Stability of periodic waves of finite amplitude on the surface of a deep fluid. J. Appl. Mech. Tech. Phys., **9**, 190–194 (1968) (originally published in Zhurnal Prikladnoi Mekhaniki i Tekhnicheskoi Fiziki,, **9**, 86–94 (1968))
193. Zakharov, V. E., Kuznetsov, E.A.: Hamiltonian formalism for nonlinear waves. Physics-Uspekhi, **40**, 1087–1116 (1997).
194. Ziglin, S. L.: Non-integrability ofa problem of a problem on the motion of four point vortices. Soviet Math. Dokl., **21**, 296–299 (1980).

Index

B. N. Shashikanth, *Dynamically Coupled Rigid Body-Fluid Flow Systems*,
https://doi.org/10.1007/978-3-030-82646-8

The manufacturer's authorised representative in the EU is Springer Nature Customer Service Centre GmbH, Europaplatz 3, 69115 Heidelberg, Germany. If you have any concerns regarding our products, please contact ProductSafety@springernature.com

Printed and bound by CPI Group (UK) Ltd, Croydon, CR0 4YY
30/09/2026
02225694-0002